Andreas Henrici

Perturbation Theory for Fermi-Pasta-Ulam Chains

Andreas Henrici

Perturbation Theory for Fermi-Pasta-Ulam Chains

A Case Study in KAM Theory

Südwestdeutscher Verlag für Hochschulschriften

Impressum/Imprint (nur für Deutschland/ only for Germany)
Bibliografische Information der Deutschen Nationalbibliothek: Die Deutsche Nationalbibliothek verzeichnet diese Publikation in der Deutschen Nationalbibliografie; detaillierte bibliografische Daten sind im Internet über http://dnb.d-nb.de abrufbar.

Alle in diesem Buch genannten Marken und Produktnamen unterliegen warenzeichen-, markenoder patentrechtlichem Schutz bzw. sind Warenzeichen oder eingetragene Warenzeichen der jeweiligen Inhaber. Die Wiedergabe von Marken, Produktnamen, Gebrauchsnamen, Handelsnamen, Warenbezeichnungen u.s.w. in diesem Werk berechtigt auch ohne besondere Kennzeichnung nicht zu der Annahme, dass solche Namen im Sinne der Warenzeichen- und Markenschutzgesetzgebung als frei zu betrachten wären und daher von jedermann benutzt werden dürften.

Verlag: Südwestdeutscher Verlag für Hochschulschriften Aktiengesellschaft & Co. KG
Dudweiler Landstr. 99, 66123 Saarbrücken, Deutschland
Telefon +49 681 37 20 271-1, Telefax +49 681 37 20 271-0, Email: info@svh-verlag.de
Zugl.: Zürich, Universität Zürich, Dissertation, 2008

Herstellung in Deutschland:
Schaltungsdienst Lange o.H.G., Berlin
Books on Demand GmbH, Norderstedt
Reha GmbH, Saarbrücken
Amazon Distribution GmbH, Leipzig
ISBN: 978-3-8381-0503-1

Imprint (only for USA, GB)
Bibliographic information published by the Deutsche Nationalbibliothek: The Deutsche Nationalbibliothek lists this publication in the Deutsche Nationalbibliografie; detailed bibliographic data are available in the Internet at http://dnb.d-nb.de.

Any brand names and product names mentioned in this book are subject to trademark, brand or patent protection and are trademarks or registered trademarks of their respective holders. The use of brand names, product names, common names, trade names, product descriptions etc. even without a particular marking in this works is in no way to be construed to mean that such names may be regarded as unrestricted in respect of trademark and brand protection legislation and could thus be used by anyone.

Publisher:
Südwestdeutscher Verlag für Hochschulschriften Aktiengesellschaft & Co. KG
Dudweiler Landstr. 99, 66123 Saarbrücken, Germany
Phone +49 681 37 20 271-1, Fax +49 681 37 20 271-0, Email: info@svh-verlag.de

Copyright © 2009 by the author and Südwestdeutscher Verlag für Hochschulschriften Aktiengesellschaft & Co. KG and licensors
All rights reserved. Saarbrücken 2009

Printed in the U.S.A.
Printed in the U.K. by (see last page)
ISBN: 978-3-8381-0503-1

Contents

Preface   v

1 **Overview and Results**   1
    1.1 Setup 1
    1.2 Periodic Chains 3
    1.3 Dirichlet Chains 7
    1.4 Geometry of the Moment Map 9
    1.5 Applications 12
    1.6 Related Work 13

2 **Theoretical Background**   15
    2.1 Hamiltonian Systems 15
    2.2 Integrable Systems 17
    2.3 Fixed Points, Birkhoff Normal Form 19
    2.4 Perturbed Integrable Systems 21

3 **Normal Form Theory**   25
    3.1 Periodic Chains 25
    3.2 Dirichlet Chains 43

4 **Nondegeneracy and Convexity**   51
    4.1 Periodic Chains 51
    4.2 Dirichlet Chains 59

5 **Foliation by the Moment Map**   65
    5.1 Foliation of $\mathcal{F}_0$ 67
    5.2 Foliation of $\mathcal{F}_{\frac{N}{4}}$ for $\frac{N}{4} \in \mathbb{Z}$ 67
    5.3 Foliation of $\mathcal{F}_k$ for $0 < k < \frac{N}{4}$ 69

6 **Discussion and Outlook**   87

A **Nonresonance Lemma**   91

B **Lemma on Symmetric Polynomials**   99

Bibliography	**103**
Index	**111**

List of Figures

1.1 Subsets of parameters (γ, r) with hyperbolic dynamics of X_γ for $N = 80$ and $k = 1, 4, 10, 19$. 11
5.1 Sets of solutions (l_1, l_2) of (5.55) for $N = 48$, $r = 1$, $\gamma = -1.35$ . . 80
5.2 Sets of solutions (l_1, l_2) of (5.55) for $N = 48$, $r = 1$, $\gamma = 0.35$ . . . 81
5.3 Sets of solutions (l_1, l_2) of (5.55) for $N = 48$, $r = 1$, $\gamma = 0.6$ . . . 81
5.4 Sets of solutions (l_1, l_2) of (5.55) for $N = 48$, $r = 0.3$, $\gamma = -2$ . . 82
5.5 Sets of solutions (l_1, l_2) of (5.55) for $N = 48$, $r = 0.3$, $\gamma = -1$ . . 82
5.6 Sets of solutions (l_1, l_2) of (5.55) for $N = 48$, $r = 0.3$, $\gamma = 3$ . . . 83

Preface

> *And certainly the atoms did not move by volition,*
> *nor did they place themselves by sharp intelligence,*
> *nor did they agree what movements to produce,*
> *but they, being many and moving about in many ways,*
> *are constantly being buffeted and given motion,*
> *and by trying every kind of combination*
> *and motion, finally they fall into the arrangements*
> *and the patterns of which the sum of things consists.*
>
> Lucretius [60] (Book I, lines 1021-1028)

The Fermi-Pasta-Ulam problem The simulations performed in the early 1950's by the physicist Enrico Fermi (1901-1954), the computer scientist John Pasta (1918-1981), and the mathematician Stanislaw Ulam (1909-1984), in a "professional configuration foreshadowing the disciplinary alliance of the future" [98], can without exaggeration be considered as having revolutionized the scientific field of nonlinear dynamics, which for about half a century had seen less progress than other areas of theoretical physics such as quantum mechanics and general relativity. Before discussing the simulations of Fermi, Pasta, and Ulam (in the sequel called "FPU") and their paradoxical outcome in greater detail, let us briefly "set the stage" by giving a quick review of the situation in the field of dynamical systems in the middle of the twentieth century.

The entire history of the theory of mechanics, and in particular dynamical systems, is also a history of mutual influence between mathematical and physical research, a process which is continuing until the present time. The formalization of mechanics at the beginning of the seventeenth century is associated with the names of Galileo Galilei (1564-1642) and Johannes Kepler (1571-1630), who discovered some of the laws of terrestrial and celestial mechanics, respectively. Isaac Newton (1643-1727) showed that the phenomena in these two areas could actually be described by the same principles, for the formulation of which he also (co-)invented calculus. The laws of mechanics were then reformulated by Joseph-Louis Lagrange (1736-1813) and William Rowan Hamilton (1805-1865), who formulated them in terms of evolutions in configuration and phase space, respectively. Hamilton's framework is still widely used to describe (classical) mechanical systems, but which also turned out to be a suitable starting point for the description of quantum-mechanical systems. Towards the end of the nineteenth century, the investigation of physical systems with a (very) large number of particles led to the new field of statistical mechanics, pioneered, among others, by Ludwig Boltzmann (1844-1906), which soon was (and still is) interwoven with probability theory, since the macroscopic description of such systems primarily consists of probabilistic statements.

On the other hand, Henri Poincaré (1854-1912) pioneered the *qualitative* study of dynamical systems, since it soon became clear that many important systems could not be solved analytically. Poincaré also developed the method of perturbations to investigate systems which could be considered as small perturbations of a system which was better understood, and it is precisely this approach that Kolmogorov used half a century later in his seminal work, to which we will return below. And even though Kolmogorov applied this method to systems which are not a priori described statistically, the main result of his theory is a statement of somewhat probabilistic nature - we will formulate it precisely in this book.

As mentioned above, in the first half of the twentieth century, despite some important contributions e.g. by George D. Birkhoff (1884-1944), the theory of dynamical systems did not evolve as rapidly as other areas of mathematical physics (we will also return to Birkhoff's work). In the study of dynamical systems, a lot of work was devoted to the notion of *ergodicity*, and most people believed in the "ergodic hypothesis", namely that arbitrarily small perturbations could turn an integrable system into an ergodic one (on each energy surface). Ironically, it was Fermi [22] himself who published a "proof" of this hypothesis, which however later turned out to be incorrect.

When the MANIAC-I computer was built in 1952 by the Theoretical Division of the Los Alamos National Laboratory, it was Fermi's ingenious idea to use it as a tool for the simulation of physical systems. Thus, together with Pasta and Ulam, he proposed to test the ergodicity hypothesis mentioned above on a comparatively simple system. Precisely, he intended to observe energy sharing among *nonlinearly* coupled rigid masses in one-dimensional chains, in the sequel called *FPU chains*, with fixed endpoints. However, instead of the expected outcome, the results would eventually turn out to be "a challenge for the foundation of physics" [14]. Let us first briefly turn to a system of *linearly* coupled masses.

The behavior of such a system, i.e. a system in which the force on each mass point depends linearly on the distance between itself and its two nearest neighbors, is completely predictable and best described in terms of normal or Fourier modes, in which the Hamiltonian (the total energy) of the system is the sum of the energies of the single modes, in other words, the system can be described as a system of *uncoupled* harmonic oscillators with no exchange of energy between different modes. When the forces between the masses are assumed to be nonlinear, as in the setup of FPU's simulation, additional coupling terms appear in the Hamiltonian, which according to the principles of statistical mechanics led Fermi to the expectation that the energy would eventually be equally distributed among the different modes (equipartition), or, as Weissert [98] puts it, "energy should march through the sequence of harmonic modes like champagne spilling down a pyramid of glasses". It was the intent of FPU's simulation to measure the rate of this expected "thermalization". The number of particles was chosen as 16, 32, or 64 (apparently related to the binary arithmetic of the computer used for the simulation), and the simulations were performed over 14'000 to 19'000 time cycles[1].

However, this thermalization was *not* observed, the energy was *not* equipartitioned among the Fourier modes. As initial condition, the entire energy was concentrated in the first mode, and it never seemed to be dispersed beyond the first few modes. More precisely, the energy seemed to oscillate between these first few modes in a "quasi-periodic" way. As stated in [18], the FPU "paradox" (as the results were henceforth called) "shows that nonlinearity is not enough to guarantee equipartition of energy". This apparent contradiction to the expectations demanded an explanation - however, at the time (1954), there was no general theory accomplishing this task, and Fermi immediately recognized the importance of the observations. Unfortunately, due to his death in November 1954, he could not contribute to resolving the paradox any more - his death also considerably delayed the publication of the surprising results. The year in which the report [23] of Fermi, Pasta, and Ulam was finally published (1955) is now widely considered as the "birth" of the FPU problem, and in the last few years numerous reviews on the history of the FPU problem have been published (see e.g. [8, 13, 26, 02]), especially at the occasion of the fiftieth centenary (2005) of the publication of the original report.

Since then, various explanations have been offered to explain the FPU paradox. Following the review articles [8] or [18], one can distinguish two types of explanations, the perturbative and the "soliton-based" approaches. In this work, we primarily follow the former approach. Besides the boundary conditions chosen by FPU (which are also known as Dirichlet boundary conditions), we also consider FPU chains with periodic boundary conditions, and here we distinguish between chains with an even and an odd number of particles, in the sequel called even periodic and odd periodic chains. The periodic case turns out to be more general from the theoretical point of view, in the sense that the theory of chains with Dirichlet boundary conditions can be treated as an application of the theory of even periodic chains.

We show that for all three types of chains and all parameter values, the FPU Hamiltonian can be approximated by an integrable system up to fourth order. In other words, *we construct a fourth-order integrable model for FPU chains*. Furthermore, in the case of the boundary conditions chosen by FPU, we show that for almost all parameter values, *the KAM theorem can be applied locally around the equilibrium point*. The KAM theorem is a result by Andrej Kolmogorov (1903-1987), Vladimir Arnol'd (*1937), and Jürgen Moser (1928-1999) asserting that the "majority" of the orbits of a slightly perturbed integrable system remains quasi-periodic, under a certain nondegeneracy condition on the

---

[1] According to Ford [26] the actual programming work was done by Mary Tsingou (*1928). As he writes, "But first, like al good scientists sporting a brand new idea, FPU began looking for someone to do the actual work." - see also [17].

frequencies of the unperturbed system. We will precisely state it in chapter 2.

Even though we thus rigorously confirm the long-standing conjecture that the KAM theorem can be applied to FPU chains for almost all parameter values, it seems unlikely that this already "explains" the FPU paradox, since it remains unclear whether the energy levels and initial conditions chosen by Fermi, Pasta, and Ulam fit into the "scheme" of the KAM theorem. In particular, since the admissible energy levels for our application of KAM to FPU appear to be becoming smaller and smaller as the number of particles tends to infinity, it seems rather unlikely that the KAM theorem is sufficient for the desired explanation of the FPU paradox.

However, we do not only rely on the KAM theorem, we also plan to numerically implement the dynamics of our integrable fourth-order approximation. If such an implementation would produce results close to FPU's original results, we think that this would be a considerable contribution towards an explanation of the FPU paradox.

Perturbative approaches were however already proposed before the KAM theorem became well-known among mathematical physicists. The first analytical approach to the FPU problem was given by Ford [24] in 1961, arguing that the missing ergodicity in the FPU was based on arithmetical properties of the unperturbed chain (here, contrary to the approach to be developed in this book, "unperturbed" means the system of uncoupled harmonic oscillators). Further work in this direction was done by Jackson [48, 49] and Ford and Waters [25].

The KAM theorem then provided new and strong theoretical support to the claim that "typical" nonlinear systems exhibit nonergodic behavior. Note that although Kolmogorov's original work dates from 1954 and the proofs of his conjecture by Arnol'd and Moser from 1962 and 1963, it took some time before their work became well known among the scientists working on the FPU problem. It is interesting to note that although "FPU" and "KAM" started more or less in the same year - 1954 -, it took at least a decade before it was realized that the latter possibly could contribute to the explanation of the former. This delay was probably caused or at least prolonged by political reasons (since "FPU" originally was a primarily American and "KAM" a primarily Soviet research area).

The first connection between FPU and KAM was made by Izrailev and Chirikov [47] in 1966. However, in this paper it was not rigorously proved that the FPU system fulfills the hypotheses of the KAM theorem, the discussion was more about the admissible relative size of the perturbation beyond which the stability asserted by the KAM theorem would "break down", leading to what today is called "strong stochasticity threshold" (more results in this direction can e.g. be found in [73, 103]). The KAM theorem then became widely known in the physics community through the article by Walker and Ford [97] where it was primarily discussed as a possible explanation of the results of the Hénon-Heiles simulation [38], a connection first observed by Gustavson [36]. Whereas after the discovery of the integrability of the Toda lattice [21, 37, 61], a special case of the system considered by FPU, it was clear that the three-particle Hénon-Heiles system could be treated as a perturbation of the (three-particle) Toda lattice, this remained unclear in the case of general FPU chains (with an arbitrary number of particles).

In the 1960's, the Japanese school around Saito, Hirooka, and Toda also became active in the research on the dynamics of nonlinear chains, in the beginning however completely unaware of the (at the time still largely American) discussion of the FPU results. These Japanese researchers even started their own numerical simulations, and according to [18] obtained results which somewhat resembled the FPU results, but they only published them after having finally learned about the FPU simulations.

In the subsequent years, the only significantly new approach towards the application of the KAM theorem to FPU chains was Nishida's idea [70] of using Birkhoff normal forms to obtain a nondegenerate integrable system, which was further elaborated by Sanders [88]; our work also follows this approach. The transformation to these Birkhoff normal forms however requires the validity of certain nonresonance conditions, which Nishida did not prove. It was only recently that Rink [82] proved Nishida's conjecture in certain special cases of the parameter values, i.e. actually carried through the transformations rigorously. In the general case, this has not been fully accomplished yet - as Weissert notes in his book [98] on (the first 20 years of) the history of the FPU problem, "Once again, although the claim was made for KAM as the probable explanation for FPU, the conditions for the theorem had not been established rigorously". As already mentioned above, it is one of the main goals of this book to rigorously establish this connection, but with the somewhat broader goal of obtaining an integrable fourth-order approximation to the FPU Hamiltonian for all types of boundary conditions.

But first let us mention another theorem providing a stability result for perturbed integrable systems, which is much less well known than the KAM theorem, namely the results of the work initiated by Nikolai N. Nekhoroshev (1946-2008) asserting stability of the motion of the perturbed system under a condition slightly stronger (and more difficult to check) than Kolmogorov's nondegeneracy, namely "steepness" of the unperturbed Hamiltonian. However, the results of this type hold for *all* sufficiently small perturbations, not just for a *majority* as in the case of the KAM theorem. Thus, this type of results is of "deterministic" rather than "probabilistic" nature, whereas there is the drawback of the slightly stronger assumptions on the unperturbed Hamiltonian. We will give a brief overview of Nekhoroshev's results (and their more recent versions) in chapter 2, and we also show that for two of the three types of boundary conditions mentioned above, in particlar the ones chosen by FPU, *the Nekhoroshev theorem can be applied to FPU chains locally around the equilibrium point.*

Another issue to be mentioned in this connection is the work of the "Italian school" on the concept of *metastability*. It can be considered as a refinement of the research thread initiated by Izrailev and Chirikov connecting FPU chains and the KAM theorem. This concept was first introduced in [31] and has been further developed in the sequel. Its main point is the observation of numerical evidence of the existence of two different time scales. As in FPU's original experiments, as initial condition the entire energy is concentrated in the lowest frequency mode. After a first time scale, one observes (up to an exponentially small tail) a constant energy distribution among the first few low frequency modes (on different energy levels), whereas complete equidistribution of energy among all modes is observed only on a second, much longer time scale. However, a thorough theoretical explanation in particular of this second time scale has apparently not yet been obtained.

The other approach towards explaining the FPU paradox consists of considering the continuum limit of the FPU chain and trying to gain insight into the discretized chain through an investigation of this continuum limit. It was Zabusky [99, 100] who first took this approach in 1962 and 1963, and then together with Kruskal in the famous paper [101] of 1965, where the discovery of "soliton" solutions of the periodic KdV equation (introduced in 1895 by Korteweg and de Vries [53]) was reported. By these soliton solutions they meant solitary wave solutions which have the property of passing through one-another and afterwards almost recovering their initial state despite the nonlinear interaction. Thus, Kruskal and Zabusky found a behavior which closely resembled the FPU observations. Even though the relation between the continuous and the discrete models were not made adequately clear at the time, the discovery of solitons subsequently led to a variety of related results. In particular, the periodic KdV equation was shown by Gardner et al. [33, 34] to be completely integrable, which together with the integrability of the Toda lattice mentioned above strongly suggested that this might explain the FPU results. However, here the problem is that even though the KdV equation can be seen as the continuum limit of the FPU and Toda chains (in a sense which we will not explain precisely, since we do not pursue this approach), it seems to be difficult to precisely explain how the properties of the continuum limit can be used to explain the behavior of the original discrete model. Until now, there is a lot of research going on in this area, which we cannot exhaustively discuss; a recent contribution is e.g. [5]. Moreover, the FPU problem also has connections with many other areas of physics such as e.g. Bose-Einstein condensation or quantum chaos - we refer to [8] for a review of some of these issues. Furthermore, there is a great amount of research on the Toda lattice, whose especially strong integrability properties lead to many important conclusions. In particular, via the "Lax formalism" [56] one can precisely consider the Toda lattice as a discrete analogue of the KdV equation, and we have also developed [41, 42, 43, 44] an analogue of Kappeler and Pöschel's normal form theory for the KdV equation [51] for the periodic Toda lattice, together with a "global'" application of the KAM and Nekhoroshev theorems, i.e. in the (almost) entire phase space, in contrast to the "local" applications of these theorems to arbitrary FPU chains presented in this book.

Concerning the Toda lattice, we emphasize that even though it is not the subject of this book, it has served as a motivation for our work on arbitrary FPU chains. In particular, it seems to be very important that *the family of dynamical systems given by the family of FPU chains contains an integrable system, namely the Toda lattice.* From a methodological point of view, this seems to be the "bottom line" of our work.

Of course, the two main threads of explanation of the FPU results, the perturbative approach and the observation of solitons in the continuum limit, do not contradict each other - as Weissert [98] notes, "it might also be said that each of these solutions describes essentially the same phenomenon from a slightly different perspective". In any case, the whole history of the FPU problem can be seen as another excellent example of the mutual influence of physical and mathematical research - in this case made possible by the availability of digital computers for scientific research.

Moreover, FPU chains are one of the most prominent examples of systems outside the realm of celestial mechanics to which the KAM theorem has been proved to be applicable. This remark holds even more in the case of the Nekhoroshev theorem - we do not know of many systems for which the conditions of the Nekhoroshev theorem have been rigorously proved to be applicable. Besides contributing to the discussion of the FPU paradox, we thus consider this work also to be a "case study" in the theory of normal forms and perturbations of integrable systems, in the sense that we have applied the theoretical tools of these areas to a specific system (or a specific class of systems) which has played an important role in the development of dynamics in general. Finally, in the case of those boundary conditions, where it turns out that we cannot directly apply the KAM theorem due to resonances making the transformation to the appropriate normal form impossible, i.e. for even periodic chains, *we perform an analysis of the geometry of the moment map of the corresponding integrable system*, which reveals surprisingly rich dynamics. We however still hope to be able to apply the KAM theorem also in this case, following an approach of San [87] and Zung [105] in simpler examples.

We again emphasize that for all types of boundary conditions we are considering, we are investigating finite-dimensional systems. The infinite-dimensional analogue, i.e. FPU chains with an infinite number of particles, is a problem which also requires different methods than the ones used in this book. There is also a great amount of work which has already been done on this topic; as an example, let us mention the series of papers by Friesecke-Pego [27, 28, 29, 30] where certain stability results for these infinite-dimensional chains are proven with methods primarily from functional analysis.

To conclude this introduction, let us make some remarks concerning the epistemological significance of the research on FPU chains. As Weissert [98] notes, the fact that the obtained results contradicted the assumptions is not the only issue to be considered: "In the history and philosophy of science, the general problem of experimental evidence that contradicts the hypothesis underlying the experiment itself is not new. However, the FPU problem is the first such case where the evidence came from the results of a simulation instead of an experiment."

As remarked in [8], the FPU problem can actually be seen as having initiated a new method of research in the physical sciences besides theoretical and experimental physics, namely a "synergetic" cooperation between phycisicts and computers, a term already used by Ulam [94]. The abundant use of this approach nowadays (also in many areas outside of physics) easily lets us forget that fifty years ago, the use of computers not just as a simple calculational device, but as a tool for studying "entire" physical systems, was revolutionary. The main idea of this type of work consists of letting theoretical predictions and numerical studies mutually influence each other, in particular of unexpected numerical results giving rise to new theoretical insights. The role of laboratory experiments was thus now taken by computer simulations.

The use of computers as a tool for the simulation of physical processes itself raises a number of epistemological issues. The main one probably is the question of why one should believe that results of a simulation actually tell us something about physical reality. Apart from the problems arising from the discretization of continuous processes necessary for a numerical implementation, one always makes some approximations and simplifying assumptions in the formulation of specific mathematical models. And there are also the philosophical issues of whether experimenters are biased in the interpretation of the outcomes of their results by certain implicit assumptions (regardless of whether the experiments are performed in the laboratory or on a computer). Since it is far beyond the scope of this book to discuss these issues, we just mention some literature where these questions are discussed, e.g. [32] or [96]. In any case, we are convinced that the FPU experiments and all the work stimulated by them are an extremely interesting "case study" for the philosophy and history of science. For instance, it would be very interesting to investigate in which sense the research on the FPU problem and its related issues can be seen as a "paradigm change" in the sense of "The Structure of Scientific Revolutions" by Kuhn [54] or how Kuhn's notions should be further developed in order to cope with the doubtless "revolutionary" research initiated by the FPU paradox. We will not discuss these questions in this work.

Summarizing our results, we prove that FPU chains with periodic and Dirichlet boundary conditions can be seen as higher order perturbations of a fourth order integrable system, and that for most types of boundary conditions and parameter values, the classical KAM theorem can be applied to these chains locally around the fixed point, i.e. for low energies. For those boundary conditions where we cannot directly apply the classical KAM thoerem, we investigate the dynamics of the moment map of the corresponding integrable system, thereby finding hyperbolic or elliptic dynamics, depending on suitably chosen bifurcation parameters.

Outline This book is a revised and slightly extended version of my thesis [39], which in turn essentially is an extended version of the papers [40] and [45] written jointly with (and under the guidance of) Thomas Kappeler. In chapter 1 we begin by presenting the formal setup of the FPU problem, thereby emphasizing several special cases of certain parameter values which are of particular interest. By distinguishing between different types of parity and boundary conditions, we arrive at three different types of chains, the odd periodic, the even periodic, and the Dirichlet chains. We then present all our results on these three types of chains. Whereas for the odd periodic and the Dirichlet chains we obtain Birkhoff normal forms up to order four and some nondegeneracy and convexity results, allowing us to apply the perturbation theory results by KAM and Nekhoroshev, for the even periodic chain we obtain a *resonant* normal form up to order four. We show that this (truncated) resonant fourth order normal form is a completely integrable system and analyze the foliation of its phase space by the moment map given by its integrals. In particular, we show that this integrable system exhibits hyperbolic dynamics, depending on bifurcation parameters.

In chapter 2, we give an overview of the theoretical background of our work. We first review the notions of a Hamiltonian system and the special case of an integrable one, explain what we mean by a Birkhoff normal form up to a certain order, and discuss different types of fixed points of Hamiltonian vector fields. Afterwards we discuss the KAM and Nekhoroshev theorems, and we mention some recent improvements of these theorems which have increased their applicability by weakening their hypotheses. In particular, we discuss the different types of nondegeneracy and convexity properties of the Hessian of the unperturbed Hamiltonian which are necessary for the application of these theorems.

The following chapter 3 is the first major part of this book. Here we perform all computations necessary for the proof of the normal form theorems on the three types of chains. These computations essentially consist of carrying through a series of transformations bringing the FPU Hamiltonian into the desired form. Even though these computations may seem unmotivated, we try to convince the reader that they are strongly inspired by our work on the periodic Toda lattice (which is not contained in this book). Nevertheless, all computations are completely self-contained. We also try to emphasize the crucial role played by the parity of the number of particles, as in the case of an even number of particles in the periodic chain there are certain fourth order resonances which do not appear in the odd case. For Dirichlet chains, we do not have to repeat all calculations of the periodic chain - it turns out that these chains can be treated as an invariant submanifold of even periodic chains, and in this special case of even periodic chains, the fourth order resonances mentioned above are no obstruction to the transformation to Birkhoff normal form of order four. Finally, in the general case (of even periodic chains) we prove the integrability of the truncated fourth order resonant normal form, which comes somewhat surprisingly, since it was previously assumed that we have this property only in certain special cases.

Chapter 4 contains the proofs of the theorems on the nondegeneracy and convexity properties of the Hessian of the odd periodic and Dirichlet chains. We do not only prove that the Hessians of these two chains are nondegenerate for almost all parameter values (in a sense to be made precise), we also derive some explicit formulas and asymptotic estimates for some of the exceptional parameter values, i.e. those where the Hessian is not nondegenerate.

In chapter 5, the third and last major part of this book, we study the geometry of the phase space of the truncated resonant normal form of the even periodic chain. We investigate how the moment map given by the integrals of the truncated fourth order resonant normal form foliate this phase space into invariant level sets, and it turns out that one can find a very rich geometry. Distinguishing between regular and critical points of the phase space, we perform various reductions depending on the rank of the differential of the moment map at these critical points in order to gain further insight into the structure of the level sets associated to these critical points. In particular, after reducing to two degrees of freedom, we find four critical points of rank zero, and we obtain a bifurcation in the space of suitably chosen parameters which determine the type (elliptic or hyperbolic) of these critical points, and we discuss the question of homo- or heteroclinic orbits. After another reduction, we briefly discuss the remaining critical points.

Finally, in chapter 6 we discuss the relevance of our results and possible directions for future research. The question of relevance is not easy to be answered and in particular depends on the results of some planned numerical work implementing some of our transformations.

Almost all proofs of the theorems presented in this book contain a lot of very explicit and sometimes tedious computations, and we admit that many of these computations may appear elementary to the experienced reader. However, it seems worthwhile to explicitly include all these computations in the book, for we consider the results of these computations to be an interesting contribution to the fifty year old discussion of the FPU paradox, and we also intend to make this book as self-contained as possible. We have only put two of these computations in appendices. Appendix A contains a complete classification of the fourth order resonances of periodic chains - it is essentially an extended and more detailed version of number theoretic results of the literature. However, it may be possible to prove (some of) these results using the integrability of the (full) periodic Toda lattice - as a proof of an algebraic result by dynamical systems methods this would also be interesting from a methodological viewpoint. Appendix B contains the proof of a combinatoral lemma necessary to prove one of the results on the asymptotic width of the "convexity interval" of the odd periodic chain. Whereas this result may be of questionable independent interest, we have included it, together with its lengthy proof, because we consider the combinatorial method of its proof to be of some interest.

Acknowledgments As already indicated, this book is a slightly extended version of my thesis [39]. Therefore, first of all, I would like to thank my advisor, Thomas Kappeler, for his advice during all the years we have been working together. It is really exceptional how much time and energy he has devoted to guiding me through my work, how many (erroneous) drafts of my papers he has meticulously examined, how much patience he brought up in guiding me through the subtleties of our research area and introducing me to new viewpoints. None of the results presented in this book would have been obtained without his help, for which I am very grateful. Especially the "geometric" parts of chapter 5 are almost exclusively due to him.

It was a great pleasure to discuss my work with Bob Rink. I have greatly profited from his previous work on the same subject, which gave my a lot of hints and insight into what can be accomplished in this area, and our various discussions in Dresden, Zürich, and Leiden (and the electronic ones) have given me deeper insight into his work, from which my work has again profited.

Many thanks also to our collaborators in Milano, in particular Luigi Galgani, Antonio Giorgilli, Dario Bambusi, Simone Paleari, and Tiziano Penati, to Jürgen Pöschel, Jochen Brüning, Nicolas Roy, and Evgeny Korotyaev for discussing with me my research. I am also very grateful to Percy Deift, who showed great interest in my work and took the time to considerably improve some of my nondegeneracy computations - that they were based on wrong assumptions is of course my fault.

I gratefully acknowledge the ongoing support of the Swiss National Science Foundation, and I thank Katrin Martin and the Südwestdeutscher Verlag für Hochschulschriften for giving me the opportunity to publish my thesis. Finally, thanks to Marie-Louise Henrici for proofreading and to Carsten Rose and Mirko Birbaumer for their help with the layout.

Zürich, August 2009 *Andreas Henrici*

Chapter 1

Overview and Results

In this first chapter we state all our results on normal forms, nondegeneracy properties, and the geometry of the moment map of one-dimensional FPU chains. First we describe the setup, and then we list our results and their applications, separated by the type of boundary conditions and the parity of the number of particles. Whereas the original FPU simulations were performed for Dirichlet boundary conditions, for the theoretical treatment it turns out to be convenient to first investigate chains with periodic boundary conditions and then treat chains with Dirichlet boundary conditions as an application of the former ones.

1.1 Setup

In this book we consider nonlinear chains with N particles of equal mass, normalized to be one, as introduced by Fermi, Pasta, and Ulam, and known as FPU chains. Let us now give the precise model of such chains. An FPU chain consists of a string of particles moving on the line or the circle interacting only with their nearest neighbors through nonlinear springs. Its Hamiltonian is given by

$$H_V = \frac{1}{2}\sum_{n=1}^{N} p_n^2 + \sum_{n=1}^{N} V(q_n - q_{n+1}), \tag{1.1}$$

where $V : \mathbb{R} \to \mathbb{R}$ is a smooth potential. Here q_n denotes the displacement of the n'th particle from its equilibrium position and p_n its momentum. The corresponding Hamiltonian equations read ($1 \leq n \leq N$)

$$\dot{q}_n = \partial_{p_n} H_V = p_n,$$
$$\dot{p}_n = -\partial_{q_n} H_V = -V'(q_n - q_{n+1}) + V'(q_{n-1} - q_n).$$

We assume either N particles and periodic boundary conditions,

$$(q_{i+N}, p_{i+N}) = (q_i, p_i) \quad \forall i \in \{0, 1\}, \tag{1.2}$$

or N' (moving) particles and Dirichlet boundary conditions, i.e. fixed endpoints,

$$q_0 = q_{N'+1} = 0. \tag{1.3}$$

We will treat Dirichlet chains as an application of periodic chains with an even number of particles, hence we first concentrate on periodic chains. Although the theory of Dirichlet chains can thus be viewed "structurally" as an application of the theory of periodic chains, they are of great independent interest, in particular because the FPU simulations were originally carried through for chains with Dirichlet boundary conditions.

Without loss of generality, the potential $V : \mathbb{R} \to \mathbb{R}$ is assumed to have a Taylor expansion at the origin of the form

$$V(x) = \kappa \left(\frac{1}{2} x^2 - \frac{\alpha}{3!} x^3 + \frac{\beta}{4!} x^4 + \dots \right), \tag{1.4}$$

where κ is the (linear) spring constant normalized to 1 and $\alpha, \beta \in \mathbb{R}$ are parameters measuring the strength of the nonlinear interaction. The minus sign in front of the parameter α in the expansion (1.4) turns out to be convenient for later computations. Substituting the expression (1.4) for V into (1.1), the corresponding expansion of H_V is given by

$$H_V = \frac{1}{2} \sum_{n=1}^{N} p_n^2 + \frac{1}{2} \sum_{n=1}^{N} (q_{n+1} - q_n)^2 + \frac{\alpha}{3!} \sum_{n=1}^{N} (q_{n+1} - q_n)^3 + \frac{\beta}{4!} \sum_{n=1}^{N} (q_{n+1} - q_n)^4 + \dots \tag{1.5}$$

For any FPU chain, the total momentum $P = \frac{1}{N} \sum_{n=1}^{N} p_n$ is an integral of motion, and therefore the center of mass $Q = \frac{1}{N} \sum_{n=1}^{N} q_n$ evolves with constant velocity. Hence any FPU chain can be viewed as a family of Hamiltonian systems of $N-1$ degrees of freedom, parametrized by the vector of initial conditions $(Q, P) \in \mathbb{R}^2$ with Hamiltonian independent of Q. In particular, for $N = 2$ any FPU chain is integrable, and hence we will concentrate on the case $N \geq 3$. Further note that for any vector $(Q, P) \in \mathbb{R}^2$, the origin in $\mathbb{R}^{2N-2}$ is an equilibrium point of the corresponding system. The momentum of such an equilibrium point is given by the constant vector $(p_1, \dots, p_N) = P(1, \dots, 1)$.

The frequencies $(\omega_k^0)_{1 \leq k \leq N-1}$ of the linearization of an arbitrary FPU chain at $(q, p) = (0, 0)$ are easily computed to be

$$\omega_k^0 = 2 \sin \frac{k\pi}{N} \tag{1.6}$$

The corresponding resonance lattice is given by

$$\left\{ l = (l_1, \dots, l_{N-1}) \in \mathbb{Z}^{N-1} \,\Big|\, \sum_{k=1}^{N-1} l_k \sin \frac{k\pi}{N} = 0 \right\}$$

and generated by the vectors $(l^{(k)})_{1 \leq k \leq N-1}$, defined by $l^{(k)} = e_k - e_{N-k}$, where $(e_i)_{1 \leq i \leq N-1}$ denotes the standard basis in $\mathbb{R}^{N-1}$.

It turns out that the properties of periodic chains near the equilibrium point strongly depend on the parity of the number N of particles. If N is odd, these chains can be transformed into Birkhoff normal form of order four, whereas if N is even, there are resonances making the analogous transformations impossible. We do not present a group-theoretic explanation of this fact. On the other hand, in the case of Dirichlet boundary conditions, the chains with an odd and an even number N' of particles behave similarly, i.e. these chains can be transformed into Birkhoff normal form of order four, independently of the parity of N'.

There are some parameter values in the expansion (1.5) which are of special importance, because of historical, phenomenological, or structural reasons. The case $\alpha = 0$ is known as the β-chain, the case $\beta = 0$ as the α-chain, and the case $\beta = \alpha^2$ is a fourth order approximation of the *Toda lattice*. The potential of the full Toda lattice is the exponential function, i.e. $V(x) = \kappa e^{-x}$, introduced by Toda [93] and extensively studied in the sequel - for an overview see e.g. [92]. It turns out that the (full) Toda lattice with periodic boundary conditions is completely integrable, which was shown independently by Flaschka [21], Hénon [37], and Manakov [61], and as mentioned in the introduction, we have also developed [41, 42, 43, 44] a normal form theory for the periodic Toda lattice in analogy to Kappeler and Pöschel's work on the periodic KdV equation [51]. In the sequel, we will however denote by Toda lattice any FPU chain with $\beta = \alpha^2$. Our results on the Toda lattice are considerably stronger than the ones on general FPU chains presented in this book; they have also served as motivation to extend the focus from the special case of the Toda lattice to the case of arbitrary FPU chains. The main consequence of this extension to a larger family of systems is that whereas our results on the Toda lattice hold in the entire (or almost the entire) phase space, our results on FPU chains hold only in a neighborhood of the origin, i.e. from a physical point of view, for low energies.

1.2 Periodic Chains

We first turn to our results on periodic chains with an odd number N of particles. For any point $(x, y) = (x_k, y_k)_{1 \leq k \leq N-1} \in \mathbb{R}^{2N-2}$ we introduce the action variables $I = (I_k)_{1 \leq k \leq N-1} \in \mathbb{R}^{N-1}$ by

$$I_k = \frac{1}{2}(x_k^2 + y_k^2). \quad (1.7)$$

Note that they are not complemented by angle variables to action-angle variables. Further we define the function $H_{\alpha,\beta} : \mathbb{R}^{N-1} \to \mathbb{R}$, given by

$$H_{\alpha,\beta}(I) := 2 \sum_{k=1}^{N-1} \sin \frac{k\pi}{N} I_k + \frac{1}{4N} \sum_{k=1}^{N-1} \left(\alpha^2 + (\beta - \alpha^2) \sin^2 \frac{k\pi}{N} \right) I_k^2$$
$$+ \frac{\beta - \alpha^2}{2N} \sum_{\substack{l \neq m \\ 1 \leq l, m \leq N-1}} \sin \frac{l\pi}{N} \sin \frac{m\pi}{N} I_l I_m. \quad (1.8)$$

Theorem 1.2.1. *Let $\alpha, \beta \in \mathbb{R}$ with $(\alpha, \beta) \neq (0,0)$. If $N \geq 3$ is odd, then any periodic FPU chain admits a Birkhoff normal form of order four. More precisely, there are canonical coordinates $(x_k, y_k)_{1 \leq k \leq N-1}$ so that the Hamiltonian of any periodic FPU chain, when expressed in these coordinates, takes the form*

$$\frac{NP^2}{2} + H_{\alpha,\beta}(I) + O(|(x,y)|^5),$$

with $H_{\alpha,\beta}(I)$ given by (1.8).

Corollary 1.2.2. *Near the equilibrium state, any periodic FPU chain with an odd number N of particles can be approximated up to order four relative to its center of mass coordinates by an integrable system of $N-1$ harmonic oscillators which are coupled at fourth order except if $\beta = \alpha^2$ (Toda lattice).*

Denote by $Q_{\alpha,\beta}$ the Hessian of $H_{\alpha,\beta}(I)$ at $I = 0$. Note that $Q_{\alpha,\beta}$ is an $(N-1) \times (N-1)$ matrix which only depends on the parameters α and β.

Theorem 1.2.3. *(i) For any given $\alpha \in \mathbb{R} \setminus \{0\}$, $\det(Q_{\alpha,\beta})$ is a polynomial in β of degree $N-1$ and has $N-1$ pairwise different real zeroes. When listed in increasing order, the zeroes $\beta_k = \beta_k(\alpha)$ satisfy*

$$0 < \beta_1 < \alpha^2, \quad 2\alpha^2 < \beta_2 < \ldots < \beta_{N-1}$$

and contain the $\frac{N-1}{2}$ distinct numbers

$$\alpha^2 \cdot \left(1 + \sin^{-2}\frac{k\pi}{N}\right) \quad \left(1 \leq k \leq \frac{N-1}{2}\right).$$

When considered as functions $\beta_k = \beta_k^{(N)}(\alpha)$ of N, the zeroes β_1 and β_2 satisfy

$$\beta_1 \to \alpha^2, \quad \beta_2 \to 2\alpha^2 \quad (N \to \infty). \tag{1.9}$$

Moreover index$(Q_{\alpha,\beta})$, defined as the number of negative eigenvalues of $Q_{\alpha,\beta}$, is given by

$$\text{index}(Q_{\alpha,\beta}) = \begin{cases} 1 & \text{for } \beta < \beta_1, \\ 0 & \text{for } \beta_1 < \beta < \beta_2, \\ N-2 & \text{for } \beta > \beta_{N-1}. \end{cases}$$

Hence, at $I = 0$, $H_{\alpha,\beta}$ is convex if and only if $\beta_1 < \beta < \beta_2$. Moreover, $H_{\alpha,\beta}$ is quasiconvex if and only if $\beta \notin [\beta_2, \beta_{N-1}]$, and directionally quasiconvex for any $\beta \in \mathbb{R}$.

(ii) For $\alpha = 0$, $\det(Q_{0,\beta})$ is a polynomial in β of degree $N-1$, and $\beta = 0$ is the only zero of $\det(Q_{0,\beta})$. It has multiplicity $N-1$, and the index of $Q_{0,\beta}$ is given by

$$\text{index}(Q_{0,\beta}) = \begin{cases} 1 & \text{for } \beta < 0, \\ N-2 & \text{for } \beta > 0. \end{cases}$$

Moreover, for any $\beta \neq 0$, $H_{0,\beta}$ is quasiconvex at $I = 0$ (and therefore also directionally quasiconvex).

1.2. PERIODIC CHAINS

Periodic FPU chains with an even number N of particles do not admit a Birkhoff normal form up to order four due to resonances except if $\beta = \alpha^2$ (Toda lattice). Applied to even periodic FPU chains, our method of analyzing odd periodic FPU chains leads to a *resonant* normal form up to order four.

Besides the variables $I = (I_k)_{1 \leq k \leq N-1}$ defined by (1.7), it turns out to be convenient to introduce the variables $M = (M_k)_{1 \leq k \leq N-1}$, $J = (J_k)_{1 \leq k \leq N-1}$, and $L = (L_k)_{1 \leq k \leq N-1}$. They are defined on $\mathbb{R}^{2N-2}$ and take values in $\mathbb{R}^{N-1}$, given by

$$M_k := \frac{1}{2}(x_k y_{N-k} - x_{N-k} y_k); \quad J_k := \frac{1}{2}(x_k x_{N-k} + y_k y_{N-k}); \qquad (1.10)$$

$$L_k := \frac{1}{2}(I_k - I_{N-k}) = \frac{1}{4}\left(x_k^2 + y_k^2\right) - \frac{1}{4}\left(x_{N-k}^2 + y_{N-k}^2\right).$$

Note that for any $1 \leq k \leq N-1$, $(M_k, J_k, L_k) = (-M_{N-k}, J_{N-k}, -L_{N-k})$, as well as

$$I_k I_{N-k} = M_k^2 + J_k^2, \qquad (1.11)$$

or

$$\left(\frac{I_k + I_{N-k}}{2}\right)^2 = M_k^2 + J_k^2 + L_k^2, \qquad (1.12)$$

i.e. M_k, J_k, L_k are the Hopf variables expressed in $x_k, y_k, x_{N-k}, y_{N-k}$. They describe the image of the Hopf map from the three-dimensional sphere of radius $\frac{1}{2}(I_k + I_{N-k})$ centered at the origin of $\mathbb{R}^4$. Further introduce

$$R_{\alpha,\beta}(J, M) := \frac{\beta - \alpha^2}{4N}\left(R(J, M) + R_{\frac{N}{4}}(J, M)\right) \qquad (1.13)$$

where

$$R(J, M) = 4 \sum_{1 \leq k < \frac{N}{4}} \sin\frac{2k\pi}{N}\left(J_k J_{\frac{N}{2}-k} - M_k M_{\frac{N}{2}-k}\right) \qquad (1.14)$$

and

$$R_{\frac{N}{4}}(J, M) = \begin{cases} J_{\frac{N}{4}}^2 - M_{\frac{N}{4}}^2 & \text{if } \frac{N}{4} \in \mathbb{N}, \\ 0 & \text{otherwise.} \end{cases} \qquad (1.15)$$

Note that for $\alpha, \beta \in \mathbb{R}$ with $\beta = \alpha^2$ (Toda lattice), the expression $R_{\alpha,\beta}$ vanishes.

Theorem 1.2.4. *Let $\alpha, \beta \in \mathbb{R}$ with $(\alpha, \beta) \neq (0, 0)$. If $N \geq 4$ is even, there are canonical coordinates $(x_k, y_k)_{1 \leq k \leq N-1}$ so that the Hamiltonian of any periodic FPU chain, when expressed in these coordinates, takes the form $H_V^{trunc}(I, J, M) + O(|(x, y)|^5)$ where*

$$H_V^{trunc}(I, J, M) = \frac{NP^2}{2} + H_{\alpha,\beta}(I) - R_{\alpha,\beta}(J, M) \qquad (1.16)$$

and where $H_{\alpha,\beta}(I)$ and $R_{\alpha,\beta}(J, M)$ are given by (1.8) and (1.13), respectively.

As already mentioned, the Hamiltonian H_V in general (i.e. if $\beta \neq \alpha^2$) cannot be transformed into Birkhoff normal form up to order four due to resonances. Nevertheless, the Hamiltonian truncated at order four, H_V^{trunc}, given by (1.16), turns out to be integrable. The form of the resonance lattice introduced above suggests that $I_k + I_{N-k}$ ($1 \leq k \leq \frac{N}{2}$) are integrals of H_V^{trunc} in involution. To find the remaining commuting integrals we express $H_{\alpha,\beta}(I)$ in terms of $I_k + I_{N-k}$ ($1 \leq k \leq \frac{N}{2}$) and a remainder term,

$$H_{\alpha,\beta}(I) = H^{(2)}(I) + H^{(4)}_{\alpha,\beta}(I) + \frac{1}{2N} \sum_{k=1}^{\frac{N}{2}-1} d_k^- I_k I_{N-k} \qquad (1.17)$$

where

$$H^{(2)}(I) = 2 \sum_{k=1}^{\frac{N}{2}-1} \sin \frac{k\pi}{N} (I_k + I_{N-k}) + 2 I_{\frac{N}{2}},$$

and

$$H^{(4)}_{\alpha,\beta}(I) = \frac{1}{4N} \sum_{k=1}^{\frac{N}{2}-1} d_k^+ (I_k + I_{N-k})^2 + \frac{\beta}{4N} I_{\frac{N}{2}}^2$$

$$+ \frac{\beta - \alpha^2}{N} I_{\frac{N}{2}} \sum_{k=1}^{\frac{N}{2}-1} \sin \frac{k\pi}{N} (I_k + I_{N-k})$$

$$+ \frac{\beta - \alpha^2}{2N} \sum_{\substack{1 \leq k,l < \frac{N}{2} \\ k \neq l}} \sin \frac{k\pi}{N} \sin \frac{l\pi}{N} (I_k + I_{N-k})(I_l + I_{N-l}),$$

with

$$d_k^\pm \equiv d_k^\pm(\alpha,\beta) := \pm \alpha^2 + (\beta - \alpha^2) \sin^2 \frac{k\pi}{N}.$$

By (1.11), the remainder term $\frac{1}{2N} \sum_{k=1}^{\frac{N}{2}-1} d_k^- I_k I_{N-k}$ in (1.17) can be written as

$$\frac{1}{2N} \left(\sum_{1 \leq k < \frac{N}{4}} \left(d_k^- (J_k^2 + M_k^2) + d_{\tilde{k}}^- (J_{\tilde{k}}^2 + M_{\tilde{k}}^2) \right) + \underbrace{d_{\frac{N}{4}}^- \left(J_{\frac{N}{4}}^2 + M_{\frac{N}{4}}^2 \right)}_{\text{only if } \frac{N}{4} \in \mathbb{N}} \right),$$

wwhere $\tilde{k} \equiv \tilde{k}(k) = \frac{N}{2} - k$. Combined with the expression (1.13) for $R_{\alpha,\beta}(J,M)$, the Hamiltonian H_V^{trunc} in (1.16) then takes the form

$$H_V^{trunc} = \frac{NP^2}{2} + H^{(2)}(I) + H^{(4)}_{\alpha,\beta}(I) + \frac{1}{2N} \sum_{1 \leq k \leq \frac{N}{4}} K_k(I,J,M), \qquad (1.18)$$

where for $1 \leq k < \frac{N}{4}$

$$K_k(I,J,M) = d_k^-(J_k^2 + M_k^2) + d_{\tilde{k}}^-(J_{\tilde{k}}^2 + M_{\tilde{k}}^2) - 2(\beta - \alpha^2) \sin \frac{2k\pi}{N} (J_k J_{\tilde{k}} - M_k M_{\tilde{k}}) \qquad (1.19)$$

1.3. DIRICHLET CHAINS

and
$$K_{\frac{N}{4}}(I,J,M) = \begin{cases} -\alpha^2 J_{\frac{N}{4}}^2 + (\beta - 2\alpha^2)M_{\frac{N}{4}}^2 & \text{if } \frac{N}{4} \in \mathbb{N}, \\ 0 & \text{otherwise.} \end{cases} \quad (1.20)$$

Note that $K_{\frac{N}{4}}(I,J,M)$ can (in the case $\frac{N}{4} \in \mathbb{N}$) also be written as

$$K_{\frac{N}{4}}(I,J,M) = -\frac{\beta - \alpha^2}{2}\left(J_{\frac{N}{4}}^2 - M_{\frac{N}{4}}^2\right) + \frac{\beta - 3\alpha^2}{2}\left(J_{\frac{N}{4}}^2 + M_{\frac{N}{4}}^2\right), \quad (1.21)$$

in analogy to (1.19).

Theorem 1.2.5. *Let $N \geq 4$ be an even integer. Then for any $\alpha, \beta \in \mathbb{R}$ with $(\alpha, \beta) \neq (0,0)$ the truncated FPU Hamiltonian H_V^{trunc} given by (1.16) is completely integrable. The following $N-1$ quantities are functionally independent integrals in involution.*

$$(I_k + I_{N-k})_{1 \leq k \leq \frac{N}{2}}, \quad (I_k + I_{\frac{N}{2}+k})_{1 \leq k < \frac{N}{4}}, \quad (K_k)_{1 \leq k \leq \frac{N}{4}}. \quad (1.22)$$

Remark: By (1.11), in the case $\beta = \alpha^2$, the formulas (1.19) and (1.21) for the integrals $(K_k(I,J,M))_{1 \leq k \leq \frac{N}{4}}$ show that they only depend on the action variables $(I_k)_{1 \leq k \leq N-1}$. This corresponds to the fact that by (1.13), the non-normal form term $R_{\alpha,\beta}(J,M)$ vanishes in this case. In other words, the periodic Toda lattice admits a Birkhoff normal form of order four regardless of the parity of N, which we already showed in [43] and which is not surprising in view of its strong integrability properties.

1.3 Dirichlet Chains

As indicated above, FPU chains with Dirichlet boundary conditions can be treated as invariant submanifolds of even periodic chains. To make this precise, consider a chain with N' ($N' \geq 3$, not necessarily even) moving particles and Hamiltonian given by

$$H_V^D = \frac{1}{2}\sum_{n=1}^{N'} p_n^2 + \sum_{n=1}^{N'} V(q_n - q_{n+1}) \quad (1.23)$$

with boundary conditions (1.3). Further, using the notation $s_k := \sin\frac{k\pi}{2N'+2}$ for any $1 \leq k \leq N'$, we define the function $H_{\alpha,\beta}^D : \mathbb{R}^{N-1} \to \mathbb{R}$ by

$$H_{\alpha,\beta}^D(I) := 2\sum_{k=1}^{N'} s_k I_k + \frac{1}{16(N'+1)}\sum_{k=1}^{N'}(\alpha^2 + 3(\beta-\alpha^2)s_k^2)I_k^2 + \underbrace{\frac{\beta-\alpha^2}{32(N'+1)}I_{\frac{N'+1}{2}}^2}_{\text{only if } \frac{N'+1}{2} \in \mathbb{N}}$$

$$+ \frac{\beta - \alpha^2}{16(N'+1)}\left(\sum_{\substack{l \neq m \\ 1 \leq l,m \leq N'}} 4s_l s_m I_l I_m - \sum_{k=1}^{N'} s_{2k} I_k I_{N'+1-k}\right). \quad (1.24)$$

Theorem 1.3.1. *Let $\alpha, \beta \in \mathbb{R}$ with $(\alpha, \beta) \neq (0,0)$. Then any FPU chain with $N' \geq 3$ moving particles and Dirichlet boundary conditions admits a Birkhoff normal form of order four, i.e. there are canonical coordinates $(x_k, y_k)_{1 \leq k \leq N'}$ so that H_V^D takes the form*

$$\frac{(N'+1)P^2}{2} + H_{\alpha,\beta}^D(I) + O(|(x,y)|^5),$$

with $H_{\alpha,\beta}^D(I)$ given by (1.24).

Corollary 1.3.2. *Near the equilibrium state, any FPU chain with N' moving particles and Dirichlet boundary conditions can be approximated up to order four by an integrable system of N' harmonic oscillators which are coupled at fourth order except if $\beta = \alpha^2$ (Toda lattice).*

Denote by $Q_{\alpha,\beta}^D$ the Hessian of $H_{\alpha,\beta}^D(I)$ at $I = 0$. Note that $Q_{\alpha,\beta}^D$ is an $N' \times N'$ matrix which only depends on the parameters α and β.

Theorem 1.3.3. *(i) For any given $\alpha \in \mathbb{R} \setminus \{0\}$, $\det(Q_{\alpha,\beta}^D)$ is a polynomial in β of degree N' and has N' real zeroes (counted with multiplicities). When listed in increasing order, the zeroes $\beta_k = \beta_k(\alpha)$ satisfy*

$$\beta_1 \leq \ldots \leq \beta_{\lceil \frac{N'+1}{2} \rceil} < \alpha^2 < \beta_{\lceil \frac{N'+3}{2} \rceil} \leq \ldots \leq \beta_{N'}.$$

Moreover index$(Q_{\alpha,\beta}^D)$, defined as the number of negative eigenvalues of $Q_{\alpha,\beta}^D$, is given by

$$\text{index}(Q_{\alpha,\beta}^D) = \begin{cases} \lceil \frac{N'+1}{2} \rceil & \text{for } \beta < \beta_1, \\ 0 & \text{for } \beta_{\lceil \frac{N'+1}{2} \rceil} < \beta < \beta_{\lceil \frac{N'+3}{2} \rceil}, \\ \lfloor \frac{N'-1}{2} \rfloor & \text{for } \beta > \beta_{N'}. \end{cases}$$

Hence, at $I = 0$, $Q_{\alpha,\beta}^D$ is convex if and only if $\beta_{\lceil \frac{N'+1}{2} \rceil} < \beta < \beta_{\lceil \frac{N'+3}{2} \rceil}$. Moreover, $H_{\alpha,\beta}^D$ is quasiconvex if and only if

$$\frac{\beta}{\alpha^2} \in \left(1 - \frac{2}{-1 + \sqrt{1 + 3\cos\frac{\pi}{2N'+2}}}, 1 + \frac{2}{1 + \sqrt{1 + 3\cos\frac{\pi}{2N'+2}}}\right),$$

and directionally quasiconvex for any $\beta \in \mathbb{R}$.

(ii) For $\alpha = 0$, $\det(Q_{0,\beta}^D)$ is a polynomial in β of degree N', and $\beta = 0$ is the only zero of $\det(Q_{0,\beta}^D)$. It has multiplicity N', and the index of $Q_{0,\beta}^D$ is given by

$$\text{index}(Q_{0,\beta}^D) = \begin{cases} \lceil \frac{N'+1}{2} \rceil & \text{for } \beta < 0, \\ \lfloor \frac{N'-1}{2} \rfloor & \text{for } \beta > 0. \end{cases}$$

Moreover, for any $\beta \neq 0$, $H_{0,\beta}^D$ is directionally quasiconvex, but not quasiconvex at $I = 0$.

1.4 Geometry of the Moment Map of the Truncated Even Periodic Chain

The integrals listed in (1.22) of the truncated FPU Hamiltonian H_V^{trunc} for an even number N of particles can be partitioned into $\lfloor \frac{N}{4} \rfloor + 1$ groups of integrals which depend only on mutually disjoint subsets of the variables $(x_k, y_k)_{1 \leq k \leq N-1}$. As a consequence, the phase space $T^*\mathbb{R}^{N-1}$ of H_V^{trunc} is the direct sum of $\lfloor \frac{N}{4} \rfloor + 1$ invariant subspaces, $T^*\mathbb{R}^{N-1} = \bigoplus_{0 \leq k \leq \frac{N}{4}} \mathcal{P}_k$, with

$$\mathcal{P}_k = \{(x_k, y_k)_{1 \leq k \leq N-1} \in T^*\mathbb{R}^{N-1} | x_j = y_j = 0 \ \forall j \notin \{k, N-k, \frac{N}{2} - k, \frac{N}{2} + k\}\},$$

and the foliations of $\mathcal{P}_0$, $\mathcal{P}_{\frac{N}{4}}$ (if $\frac{N}{4} \in \mathbb{N}$), and $\mathcal{P}_k$ for $0 < k < \frac{N}{4}$ into level sets of the integrals can be analyzed separately. We first describe the results obtained by the analysis of the foliations of $\mathcal{P}_{\frac{N}{4}} \cong T^*\mathbb{R}^2$ (if $\frac{N}{4} \in \mathbb{N}$) by the integrals $I_{\frac{N}{4}} + I_{\frac{3N}{4}}$, $K_{\frac{N}{4}}$, and then of $\mathcal{P}_k \cong T^*\mathbb{R}^4$ (for $0 < k < \frac{N}{4}$) by the integrals $I_k + I_{N-k}$, $I_{\frac{N}{2}-k} + I_{\frac{N}{2}+k}$, $I_k + I_{\frac{N}{2}+k}$, K_k (whereas $\mathcal{P}_0 \cong T^*\mathbb{R}$ is simply foliated into circles by $I_{\frac{N}{2}}$).

Note that in the case $\beta = \alpha^2$, as already remarked above, the integrals of Theorem 1.2.5 can be replaced by the action variables $I_1, \ldots, I_{N-1}$, so it remains to analyze the case $\beta \neq \alpha^2$. Instead of α and β, it turns out to be convenient to use the bifurcation parameter

$$\gamma := \frac{\alpha^2}{\alpha^2 - \beta}.$$

The geometry of the moment map $\mathcal{M} \equiv (H, K) : T^*\mathbb{R}^2 \to \mathbb{R}^2$, with $H := I_{\frac{N}{4}} + I_{\frac{3N}{4}}$ and $K := K_{\frac{N}{4}}$, can be analyzed as follows. We use the Hopf variables defined in (1.10), in which K is given by a constant multiple of $(1+\gamma)M^2 + \gamma J^2$ ($M \equiv M_{\frac{N}{4}}$, $J \equiv J_{\frac{N}{4}}$), and observe that the origin of $T^*\mathbb{R}^2$ is the only critical point of $\mathcal{M}$ of rank zero, with $\mathcal{M}^{-1}\{(0,0)\} = \{(0,0)\}$. Then we use a Hopf map to reduce the integral K to level sets of H, obtaining for the reduced vector field X_γ induced by K the formula

$$X_\gamma = \begin{cases} (-2\gamma JL, 2(1+\gamma)ML, -2MJ) & \gamma \notin \{-1, 0\}, \\ (0, L, -J) & \gamma = 0, \\ (-L, 0, M) & \gamma = -1. \end{cases}$$

One then sees that in the case $\gamma \notin \{-1, 0\}$, the reduced vector field X_γ admits six fixed points, four of which are elliptic and the other two hyperbolic fixed points connected by heteroclinic X_γ-orbits.

To analyze the foliation of $T^*\mathbb{R}^4$ by the moment map $\mathcal{M} \equiv (H_1, H_2, G, K) : T^*\mathbb{R}^4 \to \mathbb{R}^4$ with $H_1 := I_k + I_{N-k}$, $H_2 := I_{\frac{N}{2}-k} + I_{\frac{N}{2}+k}$, $G := I_k + I_{\frac{N}{2}+k}$, and $K := K_k$, we proceed similarly. We write 1, 2 for the indices k, $\frac{N}{2} - k$, again introduce the Hopf variables $(M_i, J_i, L_i)_{1 \leq i \leq 2}$ by (1.10), and, after briefly discussing the critical points of $\mathcal{M}$ of rank one and two, reduce to level sets of H_1 and H_2 through a symplectic reduction given by the product of two Hopf maps. By this procedure, we obtain a reduced moment map

$$\mathcal{M}_\gamma : \mathbb{S}^2_{r_1} \times \mathbb{S}^2_{r_2} \to \mathbb{R}^2, \ (M_i, J_i, L_i)_{1 \leq i \leq 2} \mapsto (G, K_\gamma),$$

where $(r_i)_{1 \leq i \leq 2}$ are the values of $(H_i)_{1 \leq i \leq 2}$, together with two reduced Hamiltonian vector fields Y and X_γ of G and K_γ, respectively. We show that there exist four critical points of $\mathcal{M}_\gamma$ of rank zero, namely $(M_i, J_i, L_i)_{1 \leq i \leq 2} = \varepsilon(0, 0, r_1, 0, 0, \pm r_2)$, where $\varepsilon \in \{\pm\}$. It is convenient to introduce the second bifurcation parameter

$$r := \frac{r_1}{r_2}.$$

Whereas the two points $\varepsilon(0, 0, r_1, 0, 0, -r_2)$ turn out to be elliptic for all parameter values, on the two points $\varepsilon(0, 0, r_1, 0, 0, r_2)$ we prove the following theorem:

Theorem 1.4.1. *Assume that $1 \leq k < \frac{N}{4}$, $0 < r \leq 1$, $\varepsilon \in \{\pm\}$, and $\gamma \in \mathbb{R}$. The critical point $\varepsilon(0, 0, r_1, 0, 0, r_2)$ of $\mathcal{M}_\gamma$ is a hyperbolic fixed point of the vector field X_γ if and only if*

$$\left| \left(\gamma + \sin^2 \frac{k\pi}{N}\right) \sqrt{r} + \left(\gamma + \cos^2 \frac{k\pi}{N}\right) \frac{1}{\sqrt{r}} \right| < 2 \sin \frac{2k\pi}{N}. \qquad (1.25)$$

Otherwise it is an elliptic fixed point of X_γ. If (1.25) is satisfied, the stable and unstable manifolds of $\varepsilon(0, 0, r_1, 0, 0, r_2)$ both have dimension two. In the case $r < 1$, the connected component of $\mathcal{M}_\gamma^{-1}\{\varepsilon(r_1 - r_2, 0)\}$ containing $\varepsilon(0, 0, r_1, 0, 0, r_2)$ is a 2-dimensional torus pinched at $\varepsilon(0, 0, r_1, 0, 0, r_2)$ and consists of homoclinic X_γ-orbits. In the case $r = 1$, $\mathcal{M}_\gamma^{-1}\{(0, 0)\}$ is a 2-dimensional torus pinched at the two points $\pm(0, 0, r_1, 0, 0, r_1)$, and $\mathcal{M}_\gamma^{-1}\{(0, 0)\} \setminus \{\pm(0, 0, r_1, 0, 0, r_1)\}$ consists of heteroclinic X_γ-orbits.

On the critical points of $\mathcal{M}_\gamma$ with rank one we have the following result:

Theorem 1.4.2. *Assume that $1 \leq k < \frac{N}{4}$, $0 < r \leq 1$, and $\gamma \in \mathbb{R}$. If a point $(M_i, J_i, L_i)_{1 \leq i \leq 2} \in \mathbb{S}^2_{r_1} \times \mathbb{S}^2_{r_2} \setminus \{\pm(0, 0, r_1, 0, 0, \pm r_2)\}$ is a critical point of $\mathcal{M}_\gamma$ with rank $d\mathcal{M}_\gamma = 1$, then $(M_2, L_2) = \lambda(M_1, -J_1)$ for some $\lambda \in \mathbb{R}$, and*

$$r^2(1-l_1^2)^2 l_2^2 + (1-l_2^2)^2 l_1^2 + 2r(1-l_1^2)(1-l_2^2)(2l_1 l_2 - (\sqrt{r} d_{1,\gamma} l_1 + \frac{1}{\sqrt{r}} d_{2,\gamma} l_2)^2) = 0,$$

where l_1, l_2 denote the normalized variables $l_i := \frac{L_i}{r_i} \in (0, 1)$ $(i = 1, 2)$. Given any point $(M_1, J_1, L_1) \in \mathbb{S}^2_{r_1} \setminus \{\pm(0, 0, r_1)\}$, there exist at most eight points $(M_2, J_2, L_2) \in \mathbb{S}^2_{r_2} \setminus \{\pm(0, 0, r_2)\}$ such that $(M_i, J_i, L_i)_{1 \leq i \leq 2}$ is a critical point of $\mathcal{M}_\gamma$ with rank $d\mathcal{M}_\gamma = 1$.

By another symplectic reduction, we briefly discuss the type (hyperbolic or elliptic) of these critical points of $\mathcal{M}_\gamma$ with rank one.

1.4. GEOMETRY OF THE MOMENT MAP 11

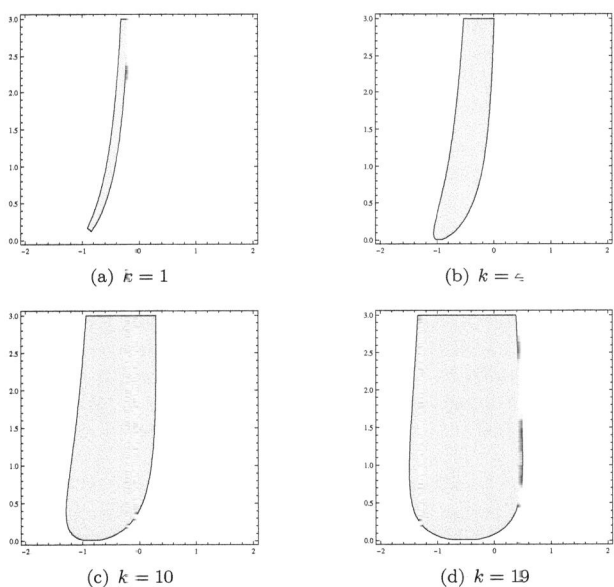

Figure 1.1: Subsets of parameters (γ, r) with hyperbolic dynamics of X_γ for $N = 80$ and $k = 1, 4, 10, 19$

1.5 Applications

We now turn to applications of our results, more precisely to those applications which could be relevant for an "explanation" of the FPU paradox described in the introduction.

The first application is that it rigorously follows that the classical KAM theorem can be applied to odd periodic and Dirichlet FPU chains locally around the equilibrium point for almost all parameter values. As mentioned in the introduction, this confirms long standing conjectures which have never been proved in full generality - we hope to close this gap with this work.

Precisely, from Theorems 1.2.1 and 1.2.3, Theorem 2.4.1 (the classical KAM theorem), and Theorems 2.4.2 or 2.4.3 (Nekhoroshev's theorem for quasiconvex and directionally quasiconvex Hamiltonians, respectively), we conclude that the following holds for odd periodic chains with Hamiltonian H_V, given by (1.1), for a real analytic potential $V(x) = \frac{1}{2}x^2 - \frac{\alpha}{3!}x^3 + \frac{\beta}{4!}x^4 + \ldots$:

Theorem 1.5.1. *For any given $\alpha \in \mathbb{R}$, the classical KAM theorem applies to odd periodic FPU chains with Hamiltonian H_V near the equilibrium point with $\beta \in \mathbb{R} \setminus \{\beta_1(\alpha), \ldots, \beta_{N-1}(\alpha)\}$ (for $\alpha = 0$, with $\beta \in \mathbb{R} \setminus \{0\}$). Moreover, Nekhoroshev's theorem applies to odd periodic chains with Hamiltonian H_V near the equilibrium point for any $(\alpha, \beta) \neq (0, 0)$.*

In the case of even periodic chains, we cannot directly apply the classical KAM or Nekhoroshev theorems because of the resonant terms in the fourth order normal form, and it is an open question whether the resonant fourth order normal for, as provided by Theorem 1.2.4, can be shown to be nondegenerate with respect to the integrals provided by Theorem 1.2.5. We think that this could be accomplished in a similar fashion as it has been done in [87, 105] for simpler examples.

Furthermore, for Dirichlet chains, in particular the type of systems as originally considered by Fermi, Pasta, and Ulam, we conclude from Theorems 1.3.1 and 1.3.3, Theorem 2.4.1, and Theorems 2.4.2 or 2.4.3, that the following holds for Dirichlet chains with Hamiltonian H_V^D, given by (1.23), for a real analytic potential $V(x) = \frac{1}{2}x^2 - \frac{\alpha}{3!}x^3 + \frac{\beta}{4!}x^4 + \ldots$:

Theorem 1.5.2. *For any given $\alpha \in \mathbb{R}$, the classical KAM theorem applies to Dirichlet FPU chains with Hamiltonian H_V near the equilibrium point with $\beta \in \mathbb{R} \setminus \{\beta_1(\alpha), \ldots, \beta_{N'}(\alpha)\}$ (for $\alpha = 0$, with $\beta \in \mathbb{R} \setminus \{0\}$). Moreover, Nekhoroshev's theorem applies to Dirichlet chains with Hamiltonian H_V near the equilibrium point for any $(\alpha, \beta) \neq (0, 0)$.*

We consider Theorems 1.5.1 and 1.5.2, the results on the application of the KAM and Nekhoroshev theorems to FPU chains, together with Theorem 1.4.1, the result of our analysis of the geometry of the moment map of the truncated Hamiltonian of even periodic chains, to be the main results of this book. In particular, for all three types of FPU chains we have constructed an *integrable approximation of the FPU Hamiltonian up to order four* (proven in Theorems 1.2.1, 1.2.4, and 1.3.1).

1.6 Related Work

The literature on FPU chains is huge, and it is almost impossible to give a complete overview; many references can also be found in the survey papers [8] and [26].

A large part of our work can be viewed as an extension to the case of arbitrary $\alpha, \beta \in \mathbb{R}$ of work on the β-chain by Rink in various papers [81, 82, 83, 85] and his thesis [84]. In particular, Theorems 1.2.1, 1.2.3, and 1.2.5 are related to results contained in [82], and Theorems 1.3.1 and 1.3.3 to results in [85]. Our study of the foliation of the phase space of the truncated even periodic chain finally is an extension of [83]. It came somewhat surprising that the integrability of truncated even periodic chains holds not only for β-chains, but for arbitrary FPU chains. Related computations for the periodic β-chain have also been performed by Poggi and Ruffo [78].

Although our work can thus be seen under this viewpoint, it is important to notice that our approach has been shaped not by the consideration of the historically important cases of the α- or the β-chain, but rather by our work on the (full) periodic Toda lattice [41, 42, 43, 44]. (We again emphasize that the proofs of our results are independent of these papers.) Thus, in our view, the most special case of an FPU chain is not the α- or the β-chain, but rather the Toda lattice, i.e. the case $\beta = \alpha^2$, with its especially strong integrability properties.

It is another question for which potentials $V(x)$ the full FPU Hamiltonian H_V (and not just the truncated Hamiltonian H_V^{trunc}) is integrable. For some contributions in this direction see [12], [35], and [80]; it seems likely that besides the (full) Toda lattice, there are not many other potentials with this property.

One of the most important open problems in the field of FPU chains is the investigation of the dynamics of these chains when the number of particles gets larger and larger - this is strongly related to the "soliton-based" approach mentioned in the introduction. It is likely that our results on FPU chains with Dirichlet boundary conditions can be used for this purpose. For recent contributions in this direction see e.g. [6, 7]. To this end, we also plan to investigate the "behavior" of our results in the limit $N \to \infty$.

Chapter 2

Theoretical Background

In this chapter we briefly explain the theoretical background of our work, namely Hamiltonian and in particular integrable systems, different types of fixed points of Hamiltonian vector fields, Birkhoff normal forms, and some theorems on perturbed integrable systems, without attempting to give a complete treatment of these notions. We largely follow the exposition in [51].

2.1 Hamiltonian Systems

Here we give a brief overview of the abstract Hamiltonian formalism, which is widely used for the mathematical description of (classical) physical systems. Let M be a smooth manifold of finite dimension without boundary, which is connected (but not necessarily compact), and let $\mathbb{F} = C^\infty(M)$. (For a review of the notion of a smooth manifold, we refer to any textbook on differential geomometry, e.g. [55].)

Definition 2.1.1. *A Poisson bracket on M is a skew-symmetric bilinear map*

$$\{\cdot,\cdot\} : \mathbb{F} \times \mathbb{F} \to \mathbb{F},$$

which satisfies the Leibniz rule,

$$\{FG, H\} = F\{G, H\} + G\{F, H\} \quad \forall F, G, H \in \mathbb{F}, \tag{2.1}$$

and the Jacobi identity,

$$\{F, \{G, H\}\} + \{G, \{H, F\}\} + \{H, \{F, G\}\} = 0 \quad \forall F, G, H \in \mathbb{F}. \tag{2.2}$$

A smooth manifold with a Poisson bracket is called a Poisson manifold.

Furthermore, a flow Φ^t on a Poisson manifold is called a *Poisson system*, if there exists a function $H \in \mathbb{F}$, the *Hamiltonian* of the system, such that

$$\dot{F} := \frac{d}{dt} F \circ \Phi^t |_{t=0} = \{F, H\} \quad \forall F \in \mathbb{F}. \tag{2.3}$$

Since the map $\mathbb{F} \to \mathbb{F}$, $F \mapsto \{F, H\}$ is a derivation, there exists a unique vector field X_H, the *Hamiltonian vector field* associated to H, such that

$$\{G, H\} = X_H G = \langle dG, X_H \rangle \quad \forall G \in \mathbb{F},$$

where $\langle \cdot, \cdot \rangle$ denotes the dual pairing between T^*M and TM. Note that due to the skew-symmetry of the Poisson bracket, we also have $\langle dG, X_H \rangle = -\langle dH, X_G \rangle$, hence we can regard X_H as a function of dH, which is linear. In other words, there exists a unique map $K : T^*M \to TM$, the *Poisson structure*, mapping each fiber T_p^*M linearly into T_pM, such that $X_H = KdH$. Since M is finite-dimensional, if the Poisson structure K is *nondegenerate*, i.e. has a trivial kernel, it must be a bijection, and we can consider the inverse $K^{-1} : TM \to T^*M$. This defines a bilinear form ν on vector fields by

$$\nu(X, Y) := \langle K^{-1}X, Y \rangle. \tag{2.4}$$

The form ν defined by (2.4) is skew-symmetric and nondegenerate (since K is a bijection). Moreover, by the Jacobi identity (2.2), $d\nu = 0$, i.e. ν is closed. In other words, ν is a *symplectic form* on M (a closed, nondegenerate 2-form), and (M, ν) is a *symplectic manifold*. Conversely, given a symplectic manifold (M, ν), one defines an isomorphism $S : TM \to T^*M$ at each point through $X \mapsto \nu \circ X$, the *symplectic structure*, and obtains for any $H \in \mathbb{F}$ the Hamiltonian vector field X_H of H by $X_H = JdH$, where $J = S^{-1} : T^*M \to TM$. This then allows to construct a Poisson bracket on M by

$$\{F, G\} = \nu(X_F, X_G) \quad \forall F, G \in \mathbb{F}.$$

Hence, in our finite-dimensional setting, symplectic forms and nondegenerate Poisson brackets are equivalent notions, and in the sequel we will not strictly distinguish between them.

An important class of diffeomorphisms of a Poisson or symplectic manifold are the diffeomorphisms preserving the underlying structure.

Definition 2.1.2. *A diffeomorphism Φ of a Poisson manifold is called* canonical, *if it preserves the Poisson bracket, i.e. if $\{F, G\} \circ \Phi = \{F \circ \Phi, G \circ \Phi\}$ for any $F, G \in \mathbb{F}$.*

A diffeomorphism Φ of a symplectic manifold is called symplectic, *if it preserves the symplectic form, i.e. if $\Phi^*\nu = \nu$.*

The standard example of a Poisson manifold is $\mathbb{R}^{2n} = \mathbb{R}^n \times \mathbb{R}^n$ for any $n \in \mathbb{N} \setminus \{0\}$ with Poisson bracket

$$\{F, G\}_0 = \langle \nabla F, J_0 \nabla G \rangle$$
$$= \langle F_q, G_p \rangle_0 - \langle F_p, G_q \rangle_0,$$

where

$$J_0 = \begin{pmatrix} 0 & I \\ -I & 0 \end{pmatrix} \tag{2.5}$$

2.2. INTEGRABLE SYSTEMS

with the n-dimensional identity matrix I, and where $\langle \cdot, \cdot \rangle_0$ denotes the standard scalar product in $\mathbb{R}^n$. For a Hamiltonian H, by (2.3), the coordinate functions $(q_i, p_i)_{1 \leq i \leq n}$ evolve by

$$\dot{q}_i = \{q_i, H\}_0 = \frac{\partial H}{\partial p_i}, \quad \dot{p}_i = \{p_i, H\}_0 = -\frac{\partial H}{\partial q_i}.$$

In terms of symplectic forms, the standard symplectic form obtained from this standard Poisson bracket is $\nu_0 = \sum_{i=1}^n dq_i \wedge dp_i$.

If the position coordinates are identified modulo 2π and thus are angular coordinates, the phase space is $\mathbb{T}^n \times \mathbb{R}^n$, where $\mathbb{T}^r = \mathbb{R}^n / 2\pi \mathbb{Z}^n$, and the coordinates are denoted by

$$(\theta, I) = (\theta_1, \ldots, \theta_n, I_1, \ldots, I_n)$$

and called *action-angle coordinates*.

2.2 Integrable Systems

A smooth non-constant function G is called an *integral* of a Hamiltonian system with Hamiltonian H, if

$$\{G, H\} = 0.$$

If a Hamiltonian system with n degrees of freedom admits n functionally independent integrals in involution, we call it *integrable*. This implies that the system can be (formally) solved for any initial data by quadratures.

Definition 2.2.1. *A Hamiltonian system on a Poisson manifold M of dimension $2n$ is called* integrable, *if its Hamiltonian H admits n functionally independent integrals $F_1, \ldots, F_n$ in involution, i.e.*

(i) $\{H, F_i\} = 0$ for $1 \leq i \leq n$ everywhere on M,

(ii) $\{F_i, F_j\} = 0$ for $1 \leq i, j \leq n$ everywhere on M, and

(iii) $dF_1 \wedge \ldots \wedge dF_n \neq 0$ on an open dense subset of M.

We call the map $F = (F_1, \ldots, F_n) : M \to \mathbb{R}^n$ the *moment map* of the integrable Hamiltonian H. The phase space M is the decomposed into *level sets* $F^{-1}(c) := \{p \in M : F(p) = c\}$ of F, in other words, it is *foliated* into the *leaves* $F^{-1}(c)$. These level sets have a particularly rich structure if the value c of F is a *critical value*, i.e. if the level set $F^{-1}(c)$ contains at least one *critical point* of F.

Definition 2.2.2. *A point $y \in M$ is a* critical point *of the moment map F of an integrable Hamiltonian system, if rank $dF(y) < n$; if rank $dF(y) = 0$, y is a* fixed point *of the system.*

If y is a critical point with rank $dF(y) > 0$, one usually performs a *symplectic reduction* to obtain a moment map of smaller dimension with fixed point y. We do not theoretically describe this procedure here, since we will describe it explicitly in our "case study" in chapter 5. A detailed theoretical treatment can be found in [16].

As an example of an integrable system, in standard action-angle coordinates $(\theta, I) \in \mathbb{T}^n \times \mathbb{R}^n$ any Hamiltonian which depends only on the action variables $I_1, \ldots, I_n$ is integrable with integrals $F_i = I_i$ for any $1 \leq i \leq n$. This example however is also typical for the case of a general integrable system, at least if one of its leaves is compact and connected, as the theorem of Liouville-Arnol'd-Jost-Mineur [4, 50] on the "semi-global" existence of action-angle coordinates and its corollaries assert. We will not cite this theorem here precisely, since we will not apply it.

In the sequel, we assume that an integrable Hamiltonian $H = H(I)$ is given in action-angle coordinates $(\theta_i, I_i)_{1 \leq i \leq n}$ with (nonvanishing) integrals $I_1, \ldots, I_n$. The equations of motion are then for any $1 \leq i \leq n$ given by

$$\dot{\theta}_i = \omega_i(I) = \frac{\partial H}{\partial I_i}(I), \quad \dot{I}_i = 0,$$

which can be immediately integrated to $\theta(t) = \theta^0 + \omega(I^0)t$, $I(t) = I^0$. The solution curves are straight lines winding around the underlying invariant torus $T_{I^0} = \mathbb{T}^n \times \{I^0\}$ with constant *frequencies*

$$\omega(I^0) = (\omega_1(I^0), \ldots, \omega_n(I^0)).$$

These tori are called *Kronecker tori*. The properties of the flow on such a Kronecker torus strongly depend on the arithmetical properties of the frequency vector ω. We distinguish between two cases:

(i) The frequencies $\omega_1, \ldots, \omega_n$ are *nonresonant*, or *rationally independent*, if

$$\langle k, \omega \rangle \neq 0 \quad \text{for all } k \in \mathbb{Z}^n \setminus \{0\}.$$

On such a torus, the orbits are dense and the flow ergodic.

(ii) The frequencies ω are *resonant*, or *rationally dependent*, if there exists some $k \in \mathbb{Z}^n \setminus \{0\}$ such that

$$\langle k, \omega \rangle = 0.$$

On such a torus, the orbits are not dense. Each orbit is dense on some lower dimensional torus.

We will see below (see (2.10)) that it is necessary to single out a sharper version of case (i), the *strongly nonresonant* frequencies.

2.3 Fixed Points, Birkhoff Normal Form

Another type of integrable systems is given in standard canonical cartesian (rectangular) coordinates $w = (q, p)$ on $\mathbb{R}^n \times \mathbb{R}^n$ with a Hamiltonian of the form

$$H = H(q_1^2 + p_1^2, \ldots, q_n^2 + p_n^2).$$

Such systems are integrable with integrals $F_i = q_i^2 + p_i^2$ ($1 \leq i \leq n$). This type of systems arises in the study of equilibria of Hamiltonian systems. One then also has the variables $r_i = \frac{q_i^2 + p_i^2}{2}$ ($1 \leq i \leq n$), which can be thought of as action coordinates, but are functions of the rectangular coordinates (q, p).

Consider now such an isolated equilibrium of a Hamiltonian system on some $2n$-dimensional symplectic manifold, i.e. an isolated fixed point of the Hamiltonian vector field. Neglecting an irrelevant additive constant the Hamiltonian, when expressed in the coordinates $w = (q, p)$ near the equilibrium with coordinates $q = 0$, $p = 0$, then has the form

$$H = \frac{1}{2} \langle Aw, w \rangle + \ldots$$

where A is the symmetric $2n \times 2n$-Hessian of H at the equilibrium point and the dots stand for terms of higher order in w. The following definition classifies the equilibrium point $w = 0$ according to the properties of the spectrum of the linearized system, $\dot{w} = J_0 A w$, where J_0 is the matrix (2.5) corresponding to the standard symplectic form $\nu_0 = \langle \cdot, J_0 \cdot \rangle$ of $\mathbb{R}^{2n}$. The classification is not complete, we only mention those types of fixed points which we encounter in our analysis of FPU chains, and it should be noted that these notions are defined for arbitrary, not necessarily Hamiltonian systems.

Definition 2.3.1. *The equilibrium point $w = 0$ is called* elliptic, *if the spectrum of $\dot{w} = J_0 A w$ is purely imaginary, i.e. if* $spec(J_0 A) = \{\pm i\lambda_1, \ldots, \pm i\lambda_n\}$ *with real numbers $\lambda_1, \ldots, \lambda_n$, and* degenerate elliptic, *if it is elliptic and at least one of these eigenvalues of $J_0 A$ is zero. It is called* hyperbolic, *if all eigenvalues of $J_0 A$ have non-zero real parts.*

If the Hamiltonian system under consideration depends on parameters, the qualitative behavior of the system might depend on the value of these parameters. In particular, with the variation of parameters, fixed points can be created or destroyed, or their stability properties can change. These qualitative changes of the behavior of the system are called *bifurcations*, the parameters causing these change are *bifurcation parameters*, and the parameter values at which these changes occur are *bifurcation points*.

If the equilibrium point $w = 0$ is elliptic, and if $spec(J_0 A)$ is simple, there exists a linear symplectic change of coordinates which brings the quadratic part of the Hamiltonian into normal form. Denoting the new coordinates by the same symbols as the old ones, one has

$$\langle Aw, w \rangle = \sum_{i=1}^{n} \lambda_i (q_i^2 + p_i^2). \tag{2.6}$$

Conversely, if the quadratic part of the Hamiltonian can be brought into the form (2.6) with $\lambda_i \in \mathbb{R}$ for $1 \leq i \leq n$, $w = 0$ is an elliptic fixed point of the Hamiltonian vector field; this follows directly from Definition 2.3.1.

Definition 2.3.2. *A Hamiltonian H is in* Birkhoff normal form up to order $m \geq 2$, *if it is of the form*

$$H = N_2 + N_4 + \ldots + N_m + H_{m+1} + \ldots, \qquad (2.7)$$

where the N_k, $2 \leq k \leq m$, are homogeneous polynomials of order k in the variables $(q_i, p_i)_{1 \leq i \leq n}$, which are actually functions of $q_1^2 + p_1^2, \ldots, q_n^2 + p_n^2$, and where $H_{m+1} + \ldots$ stands for (arbitrary) terms of order strictly greater than m. If this holds for any m, the Hamiltonian is said to be in Birkhoff normal form *and the coordinates $(q_i, p_i)_{1 \leq i \leq n}$ are referred to as* Birkhoff coordinates.

Note that if a Hamiltonian H admits a Birkhoff normal form of order m, the coefficients of the expansion (2.7) up to order m are uniquely determined, as long as the normalizing transformation is of the form id $+ \ldots$ However, the normalizing transformation is not unique.

There are well known theorems guaranteeing the existence of a Birkhoff normal form up to order m assuming that the frequencies $\lambda_1, \ldots, \lambda_n$ satisfy certain nonresonance conditions - see e.g. Theorem 4.3 in [51]. However, in the case investigated in this book, these nonresonance conditions are not satisfied for $m = 4$. We will thus not use these general theorems but rather construct the desired Birkhoff normal forms explicitly. Birkhoff's original work can be found in [9].

If the equilibrium point $w = 0$ is hyperbolic, by the Hartman-Grobman theorem (see e.g. [91]), the local phase portrait of the system near w is topologically equivalent to the phase portrait of its linearization. In such a phase portrait, one often finds *homo-* and *heteroclinic orbits*, as well as *stable* and *unstable manifolds*. These terms are however not only defined for hyperbolic fixed points, but for any fixed points of the smooth (not necessarily Hamiltonian) dynamical system $\dot{x} = f(x)$ in a domain $D \subseteq \mathbb{R}^n$. We cite the following definition from [91], an (elementary) introduction to the theory of nonlinear dynamical systems.

Definition 2.3.3. *Let x^* be a fixed point of the dynamical system $\dot{x} = f(x)$, i.e. $f(x^*) = 0$. The* stable manifold *of x^* is the set of initial conditions x_0 such that the solution $x(t)$ of $\dot{x} = f(x)$ with initial condition $x(0) = x_0$ satisfies $x(t) \to x^*$ as $t \to \infty$. Likewise, the* unstable manifold *of x^* is the set of initial conditions x_0 such that the solution $x(t)$ of $\dot{x} = f(x)$ with initial condition $x(0) = x_0$ satisfies $x(t) \to x^*$ as $t \to -\infty$.*

Moreover, trajectories of a dynamical system that start and end at the same fixed point are called homoclinic *orbits, and trajectories that start at one and end at another fixed point are called* heteroclinic *orbits.*

2.4 Perturbed Integrable Systems

Since many interesting physical systems can be viewed as perturbations of an integrable Hamiltonian system, one is interested in whether the foliation into invariant tori of the unperturbed system (as described in section 2.2) can still be found in the perturbed system, or more generally, whether the behavior of the perturbed system "resembles" the behavior of the unperturbed one. The theorems of KAM and Nekhoroshev theory give some answers to these questions.

We consider a Hamiltonian in action-angle coordinates $(\theta, I) \in \mathbb{T}^n \times D$ (where D is a bounded domain in $\mathbb{R}^n$) of the form

$$H = H_0(I) + H_\varepsilon(\theta, I) \tag{2.8}$$

with an unperturbed integrable Hamiltonian H_0 and a perturbation H_ε, which for simplicity we assume to be of the form $H_\varepsilon(\theta, I) = \varepsilon H_1(\theta, I)$. The unperturbed system is *nondegenerate*, if the frequencies vary with the actions locally in a one-to-one manner, i.e. if the frequency map

$$D \to \Omega, \quad I \mapsto \omega(I) = \frac{\partial H_0}{\partial I}(I)$$

is a local diffeomorphism everywhere in D. This is equivalent to requiring that for any $I \in D$, the frequency map $\omega(I)$ satisfies *Kolmogorov's condition*

$$\det \frac{\partial \omega}{\partial I}(I) = \det \frac{\partial^2 H_0}{\partial I^2}(I) \neq 0. \tag{2.9}$$

Whereas a dense set of tori is destroyed and a generic Hamiltonian system therefore is not integrable [62] (first results in this direction are due to Poincaré [71]), it was Kolmogorov's discovery that the *majority* of tori survives a small perturbation, namely those whose frequencies ω are not only nonresonant but *strongly nonresonant* in the sense that they satisfy a *diophantine* or *small divisor condition* of the form

$$|\langle k, \omega \rangle| \geq \frac{\alpha}{|k|^\tau} \quad \text{for all } k \in \mathbb{Z}^n \setminus \{0\}. \tag{2.10}$$

We denote for fixed $\tau > 0$ by Δ_α the set of all $\omega \in \mathbb{R}^n$ satisfying (2.10) for some given $\alpha > 0$. It can be shown that almost every $\omega \in \mathbb{R}^n$ belongs to some Δ_α in the sense that if $\tau > n - 1$, one has the Lebesgue measure estimate $\text{meas}(\Omega \setminus \Delta_\alpha) = O(\alpha)$. The parameter α in (2.10) can however not be chosen arbitrarily small for a given perturbation H_ε, it has to satisfy a condition of the form $\alpha \gg \sqrt{\varepsilon}$. For the statement of the theorem, we will fix α and thus obtain an upper bound on the parameter ε measuring the size of the perturbation. Moreover, from a bounded domain $\Omega \subset \mathbb{R}^n$ we define the subsets $\Omega_\alpha \subset \Omega$ by

$$\Omega_\alpha := \{\omega \in \Omega | \omega \in \Delta_\alpha, \text{dist}(\omega, \partial \Omega) \geq \alpha\}.$$

It can be shown that these sets Ω_α are *Cantor sets*, i.e. that they are closed, perfect, and nowhere dense, and that they satisfy the same Lebesgue measure estimate as the sets Δ_α, if the boundary of Ω is piecewise smooth.

We now state the main theorem of Kolmogorov, Arnol'd, and Moser [3, 52, 64]. For simplicity, we cite its version from [51] for systems in action-angle coordinates, and not the version for systems given in rectangular coordinates, which we will apply in this book. The latter can be found in [74] or [86] - its main condition, namely the nondegeneracy of the frequency map of the unperturbed system, is the same as in the original version for systems in action-angle coordinates.

Theorem 2.4.1. *Suppose the Hamiltonian*

$$H = H_0 + H_\varepsilon$$

is real analytic on the closure of $\mathbb{T}^n \times D$, where D is a bounded domain in $\mathbb{R}^n$. If the frequency map $I \mapsto \omega(I)$ of the integrable Hamiltonian is a local diffeomorphism everywhere in D, then there exists a constant $\delta > 0$ such that for

$$|\varepsilon| < \delta\alpha^2$$

all Kronecker tori $(\mathbb{T}^n, \omega)$ of the unperturbed system with $\omega \in \Omega_\alpha$ persist as Lagrangian tori, being only slightly deformed. Moreoever, they depend in a Lipschitz continuous way on ω and fill the phase space $\mathbb{T}^n \times D$ up to a set of measure $O(\alpha)$.

For a proof of Theorem 2.4.1, besides the original references mentioned above we refer to Pöschel's papers [74, 77].

Let us remark that several of the conditions listed in Theorem 2.4.1 can be weakened. On the one hand, neither the unperturbed Hamiltonian nor the perturbation have to be real analytic, it suffices that they are differentiable of class C^l for sufficiently large l (depending on the dimension n). On the other hand, the nondegeneracy condition can be replaced by other conditions on the frequency map $I \mapsto \omega(I)$, e.g. by isoenergetic nondegeneracy. To be precise, we call the unperturbed Hamiltonian H_0 *isoenergetically nondegenerate* in D, if for any $I \in D$,

$$\det \begin{pmatrix} \partial^2 H_0/\partial I^2 & \partial H_0/\partial I \\ \partial H_0/\partial I & 0 \end{pmatrix}(I) = \det \begin{pmatrix} \partial\omega(I)/\partial I & \omega(I) \\ \omega(I) & 0 \end{pmatrix} \neq 0. \qquad (2.11)$$

For a KAM theorem for systems satisfying (2.11) we refer to [1]. Note that the conditions (2.9) and (2.11) are independent, i.e. none of the two conditions implies the other one (see [11] for an illustration of this fact by examples).

Furthermore, it has been shown that instead of nondegeneracy or isoenergetic nondegeneracy, it suffices to require that the image of the frequency map $I \mapsto \omega(I)$ does not lie in any hyperplane in $\mathbb{R}^n$ passing through the origin (see e.g. [11] or [90]). Rüssmann then found criteria for this requirement involving terms of arbitrarily high order of the Taylor expansion of $H_0(I)$. In terms of the Taylor coefficients up to order two, Rüssmanns nondegeneracy condition means that for any $I \in D$, the columns of the Hessian of H_0 are complementary in $\mathbb{R}^n$ to $\omega(I)$, i.e. the $n \times (n+1)$-matrix

$$\bigl(\, \partial\omega(I)/\partial I \,\bigl|\, \omega(I) \,\bigr) \qquad (2.12)$$

2.4. PERTURBED INTEGRABLE SYSTEMS

has rank n. One easily sees that both (2.9) and (2.11) imply that the matrix (2.12) has rank n. There exist many further developments in KAM theory, in particular relaxations of other assumptions of the classical KAM theorem - for an extensive discussion see e.g. [57].

Even though Theorem 2.4.1 or its extensions discussed above guarantee the persistence of a majority of the invariant tori of an integrable system under a sufficiently small perturbation, the theorem is of somewhat probabilistic nature, since the tori whose frequencies do not satisfy the small divisor condition (2.10) are dense among all invariant tori of the unperturbed system (as are the tori who do satisfy (2.10)). It was Nekhoroshev [65, 66, 67] who first proved a type of result providing bounds on the variation of *all* orbits over a finite, but *exponentially long* time interval, under a slightly stronger assumption than the nondegeneracy of the KAM theorem, namely "*steepness*". In [67], he gives algebraic criteria for steepness, involving the coefficients of higher order terms of the Taylor expansion of $H_0(I)$. In [46], Il'yaschenko gives an (analytic) criterion for steepness, and in [69], Niederman gives a geometric criterion for steepness.

Because Nekhoroshev's notion of "steepness" is rather difficult to check for a given system (and the same holds for the other equivalent criteria for steepness just mentioned), we do not cite his original result here, but a more recent version requiring an even stronger assumption on the unperturbed Hamiltonian, namely *convexity* or *quasiconvexity*. Precisely, following [20], we assume that the Hamiltonian H is given in canonical rectangular coordinates $(x_i, y_i)_{1 \leq i \leq n}$ by

$$H(x,y) = \langle \omega, I \rangle + \frac{1}{2}\langle I, AI \rangle + f^{(5)}(x,y),$$

where as above $I = (\bar{I}_1, \ldots, \bar{I}_n)$ with $\bar{I}_j = \frac{1}{2}(x_j^2 + y_j^2)$ for $1 \leq j \leq n$, and $f^{(5)}$ is of order 5 in $(x_i, y_i)_{1 \leq i \leq n}$. We call the unperturbed Hamiltonian

$$H_0(I) = \langle \omega, I \rangle + \frac{1}{2}\langle I, A\bar{I} \rangle \qquad (2.13)$$

convex at $I = 0$, if the quadratic form $\frac{1}{2}\langle \xi, A\xi \rangle$ is positive or negative definite. Further, we call H_0 as in (2.13) *quasiconvex* at $I = 0$, if the restriction of $\frac{1}{2}\langle \xi, A\xi \rangle$ to the plane orthogonal to ω is either positive or negative definite, equivalently, if

$$\langle \omega, \xi \rangle = 0, \quad \frac{1}{2}\langle \xi, A\xi \rangle = 0, \quad \xi \in \mathbb{R}^n \quad \Longrightarrow \quad \xi = 0. \qquad (2.14)$$

The following theorem was first proved in [19] or [68]; in [73], Pöschel gave a new proof based on a method by Lochak [58, 59], however only in the convex case.

Theorem 2.4.2. *Suppose that the unperturbed Hamiltonian $H_0(I)$ given by (2.13) is either convex or quasiconvex at $I = 0$. Then, for $\varepsilon > 0$ sufficiently small, the estimate*

$$|I(0)| \leq \varepsilon \quad \Longrightarrow \quad |I(t)| \leq \varepsilon^a \quad \text{for} \quad |t| \leq \exp \varepsilon^{-b} \qquad (2.15)$$

holds with $a = b = \frac{1}{n}$, as well as with $a = \frac{1}{2}, b = \frac{1}{2n}$.

Let us remark that in [75], the notion of quasiconvexity is quantified in the following sense: Let D be a neighborhood of the origin. We define H_0 to be m-convex, if the inequality

$$\frac{1}{2}\langle \xi, A\xi \rangle \geq m\|\xi\|^2 \quad \forall \xi \in \mathbb{R}^n \tag{2.16}$$

holds at every point I in D. Furthermore, we define H_0 to be l, m-quasiconvex, if at every point $I \in D$ either (2.16) or

$$|\langle \omega(I), \xi \rangle| > l\|\xi\| \quad \forall \xi \in \mathbb{R}^n$$

holds. One can show that quasiconvexity in the sense (2.14) implies that this l, m-quasiconvexity condition is satisfied for some positive constants l, m.

In [20], it was proved that Nekhoroshev-type stability holds under an even weaker assumption on the unperturbed Hamiltonian than quasiconvexity. We call H_0 as in (2.13) *directionally quasiconvex* at $I = 0$, if the restriction of $\frac{1}{2}\langle I, AI \rangle$ to the plane orthogonal to ω does not vanish in the first octant, equivalently, if

$$\langle \omega, \xi \rangle = 0, \quad \frac{1}{2}\langle \xi, A\xi \rangle = 0, \quad \xi_1, \ldots, \xi_n \geq 0 \quad \Longrightarrow \quad \xi = 0. \tag{2.17}$$

Theorem 2.4.3. *Suppose that the unperturbed Hamiltonian $H_0(I)$ given by (2.13) is directionally quasiconvex at $I = 0$. Then, for $\varepsilon > 0$ sufficiently small, the estimate (2.15) is satisfied with $a = b = \frac{1}{n}$, as well as with $a = \frac{1}{2}$, $b = \frac{1}{2n}$.*

The results of KAM and Nekhoroshev theory can also be combined to obtain stronger estimates than the ones of the type (2.15). We only mention that Morbidelli and Giorgilli [63] showed that under the m-convexity assumption (2.16), tori near a KAM torus exhibit "superexponential" stability.

We finally remark that one can also prove estimates of the type (2.15) under conditions which involve the coefficients up to an order higher than two in the expansion of the unperturbed Hamiltonian in terms of the action variables $I_1, \ldots, I_n$, assuming such an expansion exists. One then obtains notions such as 3-jet nondegeneracy - see e.g. [20] and the references therein. We do not go into further detail here, since we will only apply the theorems involving the expansion of $H_0(I)$ in terms of the action variables $I_1, \ldots, I_n$ up to order two, Theorems 2.4.2 and 2.4.3.

Chapter 3

Normal Form Theory

In this chapter we explicitly carry out the necessary transformations in order to obtain the various normal forms for FPU chains with the three different types of parities and boundary conditions, as stated in Theorems 1.2.1, 1.2.4, and 1.3.1. For odd periodic and Dirichlet chains we obtain Birkhoff normal forms of order four, whereas we obtain a *resonant* normal form of order four for even periodic chains, which we then show to be completely integrable in the sense of Definition 2.2.1. In particular, in this chapter we construct integrable fourth-order approximations for all three types of FPU chains. We first consider periodic chains and then treat Dirichlet chains as invariant submanifolds of certain periodic chains.

3.1 Periodic Chains

The starting point for our computations is the expansion (1.5) of the Hamiltonian H_V of periodic FPU chains in the "physical" coordinates $(q_n, p_n)_{1 \leq n \leq N}$. The proof of our first normal form result, Theorem 1.2.1, consists of four transformations. The third and fourth of these transformations come up "naturally" if one tries to eliminate the non-normal form terms step by step, i.e. in increasing order. The first transformation - introducing relative coordinates - is an "immediate" consequence of neglecting the motion of the center of mass coordinate. The second transformation however seems unmotivated; it comes up by following our procedure of constructing Birkhoff coordinates for the periodic Toda lattice [42]. This procedure is a finite-dimensional analogue of a method used by Kappeler and Pöschel [51] for the periodic KdV equation.

First we introduce relative coordinates,

$$v_i := q_{i+1} - q_i \ (1 \leq i \leq N - 1), \qquad v_N := \frac{1}{N} \sum_{i=1}^{N} q_i, \qquad (3.1)$$

and denote by $(u_i)_{1 \leq i \leq N}$ the corresponding conjugate variables. Then, with $v = (v_i)_{1 \leq i \leq N-1}$, $(v, v_N) = Mq$ is the linear change of the coordinates $q_1, \ldots, q_N$

given by the matrix

$$M = \begin{pmatrix} -1 & 1 & 0 & \ldots & 0 \\ 0 & \ddots & \ddots & & \vdots \\ \vdots & & & & 0 \\ 0 & \ldots & 0 & -1 & 1 \\ N^{-1} & \ldots & & \ldots & N^{-1} \end{pmatrix}.$$

The variables $(u = (u_i)_{1 \leq i \leq N-1}, u_N) \in \mathbb{R}^N$ conjugate to (v, v_N) are then given by $(M^T)^{-1} p$. The matrix $(M^T)^{-1}$ can be computed to be

$$(M^T)^{-1} = \frac{1}{N} \begin{pmatrix} 1 & \ldots & & \ldots & 1 \\ 2 & \ldots & & \ldots & 2 \\ \vdots & & & & \vdots \\ \vdots & & & & \vdots \\ N & \ldots & & \ldots & N \end{pmatrix} - \begin{pmatrix} 1 & 0 & \ldots & \ldots & 0 \\ 1 & 1 & 0 & \ldots & 0 \\ \vdots & & & & \vdots \\ 1 & \ldots & & 1 & 0 \\ 0 & \ldots & & \ldots & 0 \end{pmatrix}. \quad (3.2)$$

Note that by (3.2), $u_k = kP - \sum_{j=1}^k p_j$ for any $1 \leq k \leq N-1$ (recall that $P = \frac{1}{N} \sum_{j=1}^N p_j$) and $u_N = NP$. It follows that

$$p_1 = -u_1 + P, \quad p_N = u_{N-1} + P, \quad p_k = (u_{k-1} - u_k) + P \quad (2 \leq k \leq N-1),$$

and hence

$$\frac{1}{2} \sum_{j=1}^N p_j^2 = \frac{NP^2}{2} + \frac{1}{2} \left(u_1^2 + (u_1 - u_2)^2 + \ldots + (u_{N-2} - u_{N-1})^2 + u_{N-1}^2 \right). \quad (3.3)$$

Moreover, using that $q_{N+1} - q_N = q_1 - q_N = -\sum_{k=1}^{N-1}(q_{k+1} - q_k)$ one gets for any $s \in \mathbb{N}_{\geq 1}$

$$\sum_{j=1}^N (q_{j+1} - q_j)^s = \sum_{k=1}^{N-1} v_k^s + (-1)^s \left(\sum_{k=1}^{N-1} v_k \right)^s. \quad (3.4)$$

Combining the two expressions (3.3) and (3.4) yields $H_V = \frac{NP^2}{2} + \tilde{H}_V$ with $\tilde{H}_V = H_u + H_v$, where H_u and H_v only depend on $u = (u_i)_{1 \leq i \leq N-1}$ and $v = (v_i)_{1 \leq i \leq N-1}$, respectively, and are given by

$$H_u = \frac{1}{2} \left(u_1^2 + \sum_{l=1}^{N-2} (u_{l+1} - u_l)^2 + u_{N-1}^2 \right),$$

$$H_v = \frac{1}{2} \left(\sum_{k=1}^{N-1} v_k^2 + \left(\sum_{k=1}^{N-1} v_k \right)^2 \right) + \frac{\alpha}{3!} \left(\sum_{k=1}^{N-1} v_k^3 - \left(\sum_{k=1}^{N-1} v_k \right)^3 \right)$$

$$+ \frac{\beta}{4!} \left(\sum_{k=1}^{N-1} v_k^4 + \left(\sum_{k=1}^{N-1} v_k \right)^4 \right) + O(v^5).$$

3.1. PERIODIC CHAINS

Note that for any values of α and β, the point $(v, u) = (0, 0)$ is a critical point of the Hamiltonian $\tilde{H}_V$.

The second of our four transformations brings $\tilde{H}_V = H_u + H_v$ into Birkhoff normal form up to order two. To this end, we introduce new coordinates $(\xi_k, \eta_k)_{1 \leq k \leq N-1}$, and it turns out to be convenient to use complex notation, i.e. for $1 \leq k \leq N-1$

$$\begin{cases} \zeta_k = \frac{1}{\sqrt{2}}(x_k - iy_k), \\ \zeta_{-k} = \overline{\zeta_k} = \frac{1}{\sqrt{2}}(x_k + iy_k), \end{cases} \tag{3.5}$$

where the minus sign in the definition of ζ_k is chosen so that $d\zeta_k \wedge d\zeta_{-k} = id\xi_k \wedge d\eta_k$. The vector $\zeta = (\zeta_k)_{1 \leq |k| \leq N-1}$ is an element of the space

$$\mathcal{Z} := \{z = (z_k)_{1 \leq |k| \leq N-1} \in \mathbb{C}^{2N-2} : z_{-k} = \overline{z_k} \ \forall 1 \leq k \leq N-1\}. \tag{3.6}$$

Furthermore, for the rest of this book, we introduce the notations

$$s_k := \sin\frac{k\pi}{N}, \ \lambda_k := \sqrt{|s_k|} = \sqrt{\left|\sin\frac{k\pi}{N}\right|}, \ \sigma_k := \text{sgn}(k) \quad (0 \leq |k| \leq N-1). \tag{3.7}$$

It follows immediately from (3.7) that for any $0 \leq |k| \leq N-1$, $s_k = \sigma_k \lambda_k^2$.

The proposed linear transformation $\mathcal{Z} \to \mathbb{R}^{2N-2}$, $\zeta \mapsto (v, u)$ is then defined by

$$u_1(\zeta) = \frac{1}{\sqrt{N}} \sum_{1 \leq |k| \leq N-1} \lambda_k \zeta_k, \tag{3.8}$$

$$u_{l+1}(\zeta) - u_l(\zeta) = \frac{1}{\sqrt{N}} \sum_{1 \leq |k| \leq N-1} \lambda_k e^{2\pi i l k/N} \zeta_k \quad (1 \leq l \leq N-2), \tag{3.9}$$

$$-u_{N-1}(\zeta) = \frac{1}{\sqrt{N}} \sum_{1 \leq |k| \leq N-1} \lambda_k e^{2\pi i (N-1)k/N} \zeta_k, \tag{3.10}$$

and

$$v_l(\zeta) = \frac{1}{\sqrt{N}} \sum_{1 \leq |k| \leq N-1} \lambda_k e^{2\pi i l k/N} e^{-i\pi k/N} \zeta_k \quad (1 \leq l \leq N-1). \tag{3.11}$$

Note that (3.10) is actually a consequence of (3.8) and (3.9).

Lemma 3.1.1. *The linear transformation* $\mathcal{Z} \to \mathbb{R}^{2N-2}$, $\zeta \mapsto (v, u)$, *as defined by (3.8)-(3.11), is a canonical isomorphism.*

As explained in [43], Lemma 3.1.1 follows from the construction of the Birkhoff map of the periodic Toda lattice. In order to keep this work self-contained, we now give an explicit proof.

Proof of Lemma 3.1.1. First we show that

$$\{v_l(\zeta), u_m(\zeta)\} = i\,\delta_{lm}, \tag{3.12}$$
$$\{v_l(\zeta), v_m(\zeta)\} = 0, \tag{3.13}$$
$$\{u_l(\zeta), u_m(\zeta)\} = 0, \tag{3.14}$$

for any $1 \leq l, m \leq N - 1$. Since $(v, u) = (v_i, u_i)_{1 \leq i \leq N-1}$ are canonical coordinates on $\mathbb{R}^{2N-2}$, the proof of (3.12) amounts to showing that the identity

$$\sum_{k=1}^{N-1} \left(\frac{\partial v_l}{\partial \zeta_k} \frac{\partial u_m}{\partial \zeta_{-k}} - \frac{\partial v_l}{\partial \zeta_{-k}} \frac{\partial u_m}{\partial \zeta_k} \right) = i\,\delta_{lm} \tag{3.15}$$

holds for any $1 \leq l, m \leq N - 1$. It follows from (3.8)-(3.11) that for any $1 \leq k \leq N - 1$, the derivatives appearing in (3.15) are given by

$$\frac{\partial v_l}{\partial \zeta_k} = \frac{\lambda_k}{\sqrt{N}} e^{\pi i(2l-1)k/N}, \quad \frac{\partial v_l}{\partial \zeta_{-k}} = \frac{\lambda_k}{\sqrt{N}} e^{-\pi i(2l-1)k/N},$$

$$\frac{\partial u_m}{\partial \zeta_k} = \frac{\lambda_k}{\sqrt{N}} \sum_{j=0}^{m-1} e^{2\pi i j k/N}, \quad \frac{\partial u_m}{\partial \zeta_{-k}} = \frac{\lambda_k}{\sqrt{N}} \sum_{j=0}^{m-1} e^{-2\pi i j k/N}.$$

It follows that

$$\frac{\partial v_l}{\partial \zeta_k} \frac{\partial u_m}{\partial \zeta_{-k}} - \frac{\partial v_l}{\partial \zeta_{-k}} \frac{\partial u_m}{\partial \zeta_k}$$

$$= \frac{\lambda_k^2}{N} \left(e^{\pi i(2l-1)k/N} \sum_{j=0}^{m-1} e^{-2\pi i j k/N} - e^{-\pi i(2l-1)k/N} \sum_{j=0}^{m-1} e^{2\pi i j k/N} \right)$$

$$= \frac{1}{N} \sin \frac{k\pi}{N} \sum_{j=0}^{m-1} \left(e^{\frac{\pi i k}{N}(2l-2j-1)} - e^{\frac{\pi i k}{N}(2j-2l+1)} \right)$$

$$= \frac{2i}{N} \sin \frac{k\pi}{N} \sum_{j=0}^{m-1} \sin \left(\frac{k\pi}{N}(2(l-j)-1) \right)$$

$$= \frac{i}{N} \sum_{j=0}^{m-1} \left(\cos \frac{2k\pi(1-(l-j))}{N} - \cos \frac{2k\pi(l-j)}{N} \right),$$

where for the latter identity we used that $2\sin x \sin y = \cos(x-y) - \cos(x+y)$. Taking the sum over k and changing the order of summation, we conclude that

$$\sum_{k=1}^{N-1} \left(\frac{\partial v_l}{\partial \zeta_k} \frac{\partial u_m}{\partial \zeta_{-k}} - \frac{\partial v_l}{\partial \zeta_{-k}} \frac{\partial u_m}{\partial \zeta_k} \right) = \frac{i}{N} \sum_{j=0}^{m-1} \sum_{k=1}^{N-1} \left(\cos \frac{2k\pi(1-(l-j))}{N} - \cos \frac{2k\pi(l-j)}{N} \right)$$

$$= \frac{i}{N} \sum_{j=0}^{m-1} N(\delta_{l-j,1} - \delta_{l-j,0})$$

$$= i \sum_{j=0}^{m-1} (\delta_{l,j+1} - \delta_{l,j}) = i(\delta_{lm} - \delta_{l0}) = i\delta_{lm},$$

3.1. PERIODIC CHAINS

as claimed. To prove (3.13) and (3.14) one argues similarly. From (3.12)-(3.14) it follows that the linear map $\xi \mapsto (v,u)$ is a canonical isomorphism. $\square$

We now compute $\tilde{H}_V = H_u + H_v$ in terms of these new coordinates $\zeta = (\zeta_k)_{1 \leq |k| \leq N-1}$. We compute $H_u(\zeta)$ and $H_v(\zeta)$ separately. To obtain $H_u(\zeta)$, we substitute (3.8)-(3.10) into the expression $\frac{1}{2}\left(u_1^2 + \sum_{l=1}^{N-2}(u_{l-1} - u_l)^2 + u_{N-1}^2\right)$ and obtain

$$H_u(\zeta) = \frac{1}{2N} \sum_{l=0}^{N-1} \left(\sum_{1 \leq |k| \leq N-1} \lambda_k e^{2\pi i l k/N} \zeta_k\right)^2$$

$$= \frac{1}{2N} \sum_{1 \leq |k|, |k'| \leq N-1} \lambda_k \lambda_{k'} \left(\sum_{l=0}^{N-1} e^{2\pi i l(k+k')/N}\right) \zeta_k \zeta_{k'}.$$

Using that

$$\sum_{l=0}^{N-1} e^{2\pi i l k/N} = N\delta_{k0}, \quad \lambda_k = \lambda_{-k} \text{ for any } 1 \leq |k| \leq N-1, \tag{3.16}$$

we get

$$H_u(\zeta) = \sum_{k=1}^{N-1} \sin\frac{k\pi}{N} \zeta_k \zeta_{-k}.$$

Before computing $H_v(\zeta)$, let us simplify its expansion in terms of the variables $(v_k)_{1 \leq k \leq N-1}$. To this end we v_0 by the expression on the right hand side of (3.11) evaluated at $l = 0$. Note that

$$\sum_{l=0}^{N-1} v_l = \frac{1}{\sqrt{N}} \sum_{1 \leq |k| \leq N-1} \lambda_k \zeta_k e^{-i\pi k/N} \left(\sum_{l=0}^{N-1} e^{2\pi i l k/N}\right) = 0.$$

It follows that $\sum_{l=1}^{N-1} v_l = -v_0$, and therefore we have

$$H_v = \sum_{l=0}^{N-1}\left(\frac{1}{2}v_l^2 + \frac{\alpha}{3!}v_l^3 + \frac{\beta}{4!}v_l^4\right) + O(|v|^5).$$

Substituting the expression (3.11) for v_l in the quadratic term of this expansion, we obtain

$$\frac{1}{2}\sum_{l=0}^{N-1} v_l^2 = \frac{1}{2N} \sum_{1 \leq |k|, |k'| \leq N-1} \lambda_k \lambda_{k'} \left(\sum_{l=0}^{N-1} e^{2\pi i l(k+k')/N}\right) e^{-i\pi(k+k')/N} \zeta_k \zeta_{k'}$$

$$= \sum_{k=1}^{N-1} \sin\frac{k\pi}{N} \zeta_k \zeta_{-k},$$

where we again used (3.16).

The terms of third and fourth order in H_v are treated similarly. Combining the above computations leads to

$$H_u(\zeta) = \frac{1}{2}G_2(\zeta),$$
$$H_v(\zeta) = \frac{1}{2}G_2(\zeta) + \alpha G_3(\zeta) + \beta G_4(\zeta) + O(\zeta^5),$$

where

$$G_2 := 2\sum_{k=1}^{N-1} s_k \zeta_k \zeta_{-k}, \tag{3.17}$$

$$G_3 := \frac{1}{6\sqrt{N}} \sum_{(k,k',k'')\in K_3} (-1)^{(k+k'+k'')/N} \lambda_k \lambda_{k'} \lambda_{k''} \zeta_k \zeta_{k'} \zeta_{k''}, \tag{3.18}$$

$$G_4 := \frac{1}{24N} \sum_{(k,k',k'',k''')\in K_4} (-1)^{(k+k'+k''+k''')/N} \lambda_k \lambda_{k'} \lambda_{k''} \lambda_{k'''} \zeta_k \zeta_{k'} \zeta_{k''} \zeta_{k'''}, \tag{3.19}$$

with

$$K_3 := \{(k, k', k'') \in \mathbb{Z}^3 : 1 \leq |k|, |k'|, |k''| \leq N-1$$
$$\text{and } k + k' + k'' \equiv 0 \bmod N\} \tag{3.20}$$

and

$$K_4 := \{(k, k', k'', k''') \in \mathbb{Z}^4 : 1 \leq |k|, |k'|, |k''|, |k'''| \leq N-1$$
$$\text{and } k + k' + k'' + k''' \equiv 0 \bmod N\}. \tag{3.21}$$

Note that G_2, G_3, and G_4 are *independent* of α and β. In particular they already appeared in [43] when we computed the Birkhoff normal form of the periodic Toda lattice.

Summarizing the results of the first two transformations we have that

$$\tilde{H}_V(\zeta) = G_2(\zeta) + \alpha G_3(\zeta) + \beta G_4(\zeta) + O(\zeta^5) \tag{3.22}$$

is in Birkhoff normal form up to order two. As a consequence (recall the discussion after the definition (2.3.1) of elliptic fixed points), $\zeta = 0$ is an elliptic fixed point of the Hamiltonian $\tilde{H}_V$. We will later denote by Ψ_0 the transformation between the physical coordinates (q, p) and these variables ζ in which the FPU Hamiltonian is of the form (3.22).

For the remaining transformations, we follow a standard procedure, see e.g. section 14 in [51]. The phase space $\mathcal{Z}$, defined in (3.6), is endowed with the Poisson bracket

$$\{F, G\} = i \sum_{1 \leq |k| \leq N-1} \sigma_k \frac{\partial F}{\partial \zeta_k} \frac{\partial G}{\partial \zeta_{-k}}.$$

The Hamiltonian vector field X_F associated to any given Hamiltonian F is given by $X_F = i \sum_{1 \leq |k| \leq N-1} \sigma_k \frac{\partial F}{\partial \zeta_{-k}} \frac{\partial}{\partial \zeta_k}$. With the next canonical transformation we

3.1. PERIODIC CHAINS

want to eliminate the third order term αG_3 in $\tilde{H}_V(\zeta)$. To this end, we use a by now standard procedure to construct such a canonical transformation on the phase space $\mathcal{Z}$, namely the time-1-map $\Psi_1 := X_{\alpha F_3}^t|_{t=1}$ of the flow $X_{\alpha F_3}^t$ of a real analytic Hamiltonian αF_3 which is a homogeneous polynomial in $(\zeta_k)_{1 \leq |k| \leq N-1}$ of degree three and solves the homological equation

$$\{G_2, \alpha F_3\} + \alpha G_3 = 0. \tag{3.23}$$

To simplify the notation we momentarily write F instead of αF_3 and H instead of $\tilde{H}_V$. Assuming for the moment that (3.23) can be solved and that X_F^t is defined for $0 \leq t \leq 1$ in some neighbourhood of the origin in $\mathcal{Z}$, we can use Taylor's formula to expand $H \circ X_F^t$ around $t = 0$,

$$H \circ X_F^t = H \circ X_F^0 + \int_0^t \frac{d}{ds}(H \circ X_F^s) ds$$

$$= H + \int_0^t \{H, F\} \circ X_F^s \, ds$$

$$= H + t\{H, F\} + \int_0^t ds \int_0^s ds' \frac{d}{ds'}(\{H, F\} \circ X_F^{s'})$$

$$= H + t\{H, F\} + \int_0^t (t - s)\{\{H, F\}, F\} \circ X_F^s \, ds. \tag{3.24}$$

When evaluating this expression at $t = 1$, we get

$$H \circ \Psi_1 = G_2 + \{G_2, F\} + \int_0^1 (1 - t)\{\{G_2, F\}, F\} \circ X_F^t dt$$

$$+ \alpha G_3 + \int_0^1 \{\alpha G_3, F\} \circ X_F^t \, dt + \beta G_4 - O(\zeta^5).$$

Using that $\{G_2, F\} + \alpha G_3 = 0$, the latter expression is simplified and we obtain

$$H \circ \Psi_1 = G_2 + \int_0^1 t\{\alpha G_3, F\} \circ X_F^t \, dt + \beta G_4 + O(\zeta^5).$$

Integrating by parts once more and taking into account that $F \equiv \alpha F_3$ is homogeneous of degree three one obtains, in view of (3.24),

$$\tilde{H}_V \circ \Psi_1 = G_2 + \frac{1}{2}\{\alpha G_3, \alpha F_3\} + \beta G_4 + O(\zeta^5). \tag{3.25}$$

Note that $\{\alpha G_3, \alpha F_3\}$ is homogeneous of order four, i.e. the third-order terms are eliminated, as desired. It remains to solve the homological equation (3.23). Since G_3 contains only monomials with $(k, k', k'') \in K_3$ (see (3.20) for the definition of the index set K_3), also F_3 need only contain such monomials, i.e. we make the ansatz

$$F_3 = \sum_{(k,k',k'') \in K_3} F^{(3)}_{kk'k''} \zeta_k \zeta_{k'} \zeta_{k''}, \tag{3.26}$$

which leads to

$$\{G_2, F_3\} = 2i \sum_{1 \leq |k| \leq N-1} s_k \zeta_{-k} \frac{\partial F_3}{\partial \zeta_{-k}}$$

$$= -2i \sum_{(k,k',k'') \in K_3} (s_k + s_{k'} + s_{k''}) F^{(3)}_{kk'k''} \zeta_k \zeta_{k'} \zeta_{k''}. \quad (3.27)$$

We cite the following result from Beukers and Rink (cf. [82, 85]):

Lemma 3.1.2. For any $(k, k', k'') \in K_3$,

$$s_k + s_{k'} + s_{k''} \neq 0.$$

Let us remark that Lemma 3.1.2 also follows from the integrability of the Toda lattice (cf. [43]). For completeness, we include the self-contained proof due to Beukers and Rink.

Proof. Suppose that the index triple $(k, k', k'') \in K_3$ satisfies $s_k + s_{k'} + s_{k''} = 0$. It follows from $k + k' + k'' \equiv 0 \mod N$ that either $s_{k''} = -s_{k+k'}$ or $s_{k''} = s_{k+k'}$, according to whether $k + k' + k'' \equiv 0 \mod 2N$ or $k + k' + k'' \equiv N \mod 2N$.

In the first case, it follows that

$$\sin \frac{k\pi}{N} + \sin \frac{k'\pi}{N} - \sin\left(\frac{k\pi}{N} + \frac{k'\pi}{N}\right) = 0. \quad (3.28)$$

Multiplying by $2i$ and setting $x := e^{\frac{ik\pi}{N}}$ and $y := e^{\frac{ik'\pi}{N}}$, one can rewrite (3.28) as

$$0 = x - \frac{1}{x} + y - \frac{1}{y} - xy + \frac{1}{xy} = (1-x)(1-y)(1-xy)\frac{1}{xy}. \quad (3.29)$$

It follows that any solution of (3.29) contradicts the assumption $(k, k', k'') \in K_3$, in particular $1 \leq |k|, |k'|, |k''| \leq N - 1$. Indeed, all possible solutions of (3.29) contradict this assumption, namely solutions with $x = 1$ (i.e. $k \equiv 0 \mod 2N$), $y = 1$ (i.e. $k' \equiv 0 \mod 2N$), or $xy = 1$ (i.e. $k + k' \equiv 0 \mod 2N$ and thus $k'' \equiv 0 \mod 2N$).

In the second case, instead of (3.28) we have

$$\sin \frac{k\pi}{N} + \sin \frac{k'\pi}{N} + \sin\left(\frac{k\pi}{N} + \frac{k'\pi}{N}\right) = 0. \quad (3.30)$$

With x, y as above, it now follows from (3.30) that

$$0 = x - \frac{1}{x} + y - \frac{1}{y} + xy - \frac{1}{xy} = -(1+x)(1+y)(1-xy)\frac{1}{xy}.$$

Again we conclude that any solution of (3.30) contradicts the assumption $1 \leq |k|, |k'|, |k''| \leq N - 1$, namely solutions with $x = -1$ (i.e. $k \equiv N \mod 2N$), $y = -1$ (i.e. $k' \equiv N \mod 2N$), or $xy = 1$ (i.e. $k + k' \equiv 0 \mod 2N$ and thus $k'' \equiv N \mod 2N$). □

3.1. PERIODIC CHAINS

By Lemma 3.1.2, one can define F_3 by

$$iF^{(3)}_{kk'k''} := \begin{cases} \dfrac{G^{(3)}_{kk'k''}}{2(s_k+s_{k'}+s_{k''})} & (k,k',k'') \in K_3, \\ 0 & \text{otherwise.} \end{cases}$$

Then it follows from (3.27) that the homological equation (3.28) is satisfied, i.e. $\{G_2, \alpha F_3\} + \alpha G_3 = 0$. Written more explicitly, the nonzero coefficients of F_3 are

$$iF^{(3)}_{kk'k''} = \frac{(-1)^{(k+k'+k'')/N}}{6\sqrt{N}} \frac{\sqrt{|\sin\frac{k\pi}{N}\sin\frac{k'\pi}{N}\sin\frac{k''\pi}{N}|}}{2\left(\sin\frac{k\pi}{N}+\sin\frac{k'\pi}{N}+\sin\frac{k''\pi}{N}\right)}. \tag{3.31}$$

In the next step we normalize the fourth order term $\beta G_4 + \frac{\alpha^2}{2}\{G_3, F_3\}$ in (3.25). We decompose this sum into its contribution to the Birkhoff normal form and the rest, to be transformed away by our last transformation. Computing $\{G_3, F_3\}$ in a more explicit form, we obtain by (3.18) and (3.31)

$$\{G_3, F_3\} = i\sum_{1\le |k|\le N-1} \sigma_k \frac{\partial G_3}{\partial \zeta_k}\frac{\partial F_3}{\partial \zeta_{-k}} = \sum_{1\le |k|\le N-1} \sigma_k \frac{\partial G_3}{\partial \zeta_k}\frac{\partial (iF_3)}{\partial \zeta_{-k}}$$

$$= \frac{1}{36N}\sum_{1\le |k|\le N-1}\sigma_k \left(3\sum_{\substack{1\le |l|,|m|\le N-1, r\in\mathbb{Z} \\ l+m=-k+rN}}(-1)^r \lambda_k\lambda_l\lambda_m \zeta_l\zeta_m\right)$$

$$\cdot \left(3\sum_{\substack{1\le |l'|,|m'|\le N-1, r'\in\mathbb{Z} \\ l'+m'=k+r'N}}(-1)^{r'}\frac{\lambda_k\lambda_{l'}\lambda_{m'}}{2(s_{-k}+s_{l'}+s_{m'})}\zeta_{l'}\zeta_{m'}\right)$$

$$= \frac{1}{8N}\sum_{1\le |k|\le N-1}\sum_{\substack{1\le |l|,|m|,|l'|,|m'|\le N-1 \\ r,r'\in\mathbb{Z} \\ l+m-rN=-k \\ l'+m'-r'N=k}}(-1)^{r+r'}\frac{s_k\lambda_l\lambda_m\lambda_{l'}\lambda_{m'}}{s_{-k}+s_{l'}+s_{m'}}\zeta_l\zeta_m\zeta_{l'}\zeta_{m'},$$

where for the latter equality we used that $\sigma_k\lambda_k^2 = \varepsilon_k$. Setting

$$\varepsilon_{lml'm'} := \frac{l+m+l'+m'}{N} = r+r', \tag{3.32}$$

one then gets

$$\{G_3, F_3\} = \frac{1}{8N}\sum_{1\le |k|\le N-1}\sum_{\substack{l+m\equiv -k \bmod N \\ l'+m'\equiv k \bmod N}}(-1)^{\varepsilon_{lml'm'}}\frac{\lambda_l\lambda_m\lambda_{l'}\lambda_{m'}}{-1+(s_{l'}+s_{m'})/s_k}\zeta_l\zeta_m\zeta_{l'}\zeta_{m'}$$

$$= \frac{1}{8N}\sum_{k=1}^{N-1}\sum_{\substack{l+m\equiv -k \bmod N \\ l'+m'\equiv k \bmod N}}(-1)^{\varepsilon_{lml'm'}}\frac{\lambda_l\lambda_m\lambda_{l'}\lambda_{m'}}{-1+(s_{l'}+s_{m'})/\varepsilon_k}\zeta_l\zeta_m\zeta_{l'}\zeta_{m'}$$

$$+ \frac{1}{8N}\sum_{k=1}^{N-1}\sum_{\substack{l+m\equiv k \bmod N \\ l'+m'\equiv -k \bmod N}}(-1)^{\varepsilon_{lml'm'}}\frac{\lambda_l\lambda_m\lambda_{l'}\lambda_m}{-1-(s_{l'}+s_{m'})/s_k}\zeta_l\zeta_m\zeta_{l'}\zeta_{m'}$$

$$= \frac{1}{8N} \sum_{k=1}^{N-1} \sum_{\substack{l+m \equiv -k \bmod N \\ l'+m' \equiv k \bmod N}} (-1)^{\varepsilon_{lml'm'}} \lambda_l \lambda_m \lambda_{l'} \lambda_{m'}$$

$$\cdot \left(\frac{1}{-1 + (s_{l'} + s_{m'})/s_k} + \frac{1}{-1 - (s_l + s_m)/s_k} \right) \zeta_l \zeta_m \zeta_{l'} \zeta_{m'}.$$

Note that for $k = l' + m' + r'N$ with $1 \leq k \leq N-1$ and $r' \in \mathbb{Z}$ we have

$$s_k = |s_{l'+m'}|.$$

For any $(l, m, l', m') \in K_4$ introduce[1]

$$c_{lml'm'} = \begin{cases} \dfrac{1}{-1 + \frac{s_{l'} + s_{m'}}{|s_{l'+m'}|}} - \dfrac{1}{1 + \frac{s_l + s_m}{|s_{l+m}|}} & \text{if } l + m \not\equiv 0 \bmod N, \\ 0 & \text{otherwise.} \end{cases} \quad (3.33)$$

We then obtain

$$\frac{\alpha^2}{2} \{G_3, F_3\} = \frac{\alpha^2}{16N} \sum_{(l,m,l',m') \in K_4} (-1)^{\varepsilon_{lml'm'}} c_{lml'm'} \lambda_l \lambda_m \lambda_{l'} \lambda_{m'} \zeta_l \zeta_m \zeta_{l'} \zeta_{m'}. \quad (3.34)$$

Combined with formula (3.19) for G_4, the quantity $\beta G_4 + \frac{\alpha^2}{2}\{G_3, F_3\}$ can hence be written as

$$\frac{1}{24N} \sum_{(k,k',k'',k''') \in K_4} (-1)^{\varepsilon_{kk'k''k'''}} \left(\beta + \frac{3\alpha^2}{2} c_{kk'k''k'''} \right) \lambda_k \lambda_{k'} \lambda_{k''} \lambda_{k'''} \zeta_k \zeta_{k'} \zeta_{k''} \zeta_{k'''}. \quad (3.35)$$

We now decompose (3.35) into its contribution to the Birkhoff normal form of H_V and the rest, and we denote by π_N the projection onto the former one, whereas the latter one will be (partially) transformed away by another transformation Ψ_2.

Lemma 3.1.3. *The normal form part of $\beta G_4 + \frac{\alpha^2}{2}\{G_3, F_3\}$ is given by*

$$\pi_N \left(\beta G_4 + \frac{\alpha^2}{2} \{G_3, F_3\} \right)$$
$$= \frac{1}{4N} \left(\sum_{l=1}^{N-1} (\alpha^2 + (\beta - \alpha^2) s_l^2) |\zeta_l|^4 + 2 \sum_{1 \leq l \neq m \leq N-1} (\beta - \alpha^2) s_l \, s_m |\zeta_l|^2 |\zeta_m|^2 \right). \quad (3.36)$$

Proof. The indices k, k', k'', k''' of the terms in $\beta G_4 + \frac{\alpha^2}{2}\{G_3, F_3\}$ contributing to the normal form satisfy $(k, k', k'', k''') \in K_4^N$, where

$$K_4^N := \{(k, k', k'', k''') \in K_4 | \exists 1 \leq l \leq m \leq N-1 \text{ such that}$$
$$\{k, k', k'', k'''\} = \{l, -l, m, -m\}\}. \quad (3.37)$$

---

[1] To keep the formula for $c_{lml'm'}$ as simple as possible we do not symmetrize the coefficients $c_{lml'm'}$.

3.1. PERIODIC CHAINS

In the case $l = m$, $\{l, -l, l, -l\}$ in (3.37) is viewed as a multiset, i.e. set-like object whose two elements l and $-l$ each have multiplicity two. Note that by (3.32) and (3.37), for any $(k, k', k'', k''') \in K_4^N$, we have $(-1)^{\varepsilon_{kk'k''k'''}} = 1$.

We investigate $\pi_N(\beta G_4)$ and $\pi_N(\frac{\alpha^2}{2}\{G_3, F_3\})$ separately (π_N is obviously a linear map). Let us start with $\pi_N(\beta G_4)$. We distinguish the cases $l = m$ and $l \neq m$ in K_4^N. For $l = m$, there are $\binom{4}{2} = 6$ distinct permutations of (k, k', k'', k''') in K_4^N, whereas for $l \neq m$, all $4! = 24$ permutations of $(l, m, -l, -m)$ are distinct. Hence we have

$$\pi_N(\beta G_4) = \frac{\beta}{24N}\left(6\sum_{l=1}^{N-1} s_l^2|\zeta_l|^4 + 24\sum_{1\leq l < m \leq N-1} s_l\, s_m|\zeta_l|^2|\zeta_m|^2\right)$$

$$= \frac{\beta}{4N}\left(\sum_{l=1}^{N-1} s_l^2|\zeta_l|^4 + 2\sum_{1\leq l\neq m \leq N-1} \varepsilon_l\, s_m|\zeta_l|^2|\zeta_m|^2\right). \quad (3.38)$$

Now let us compute $\pi_N(\frac{\alpha^2}{2}\{G_3, F_3\}) = \alpha^2 \pi_N(\frac{1}{2}\{G_3, F_3\})$. We have to single out the matches of the normal form condition (3.37) on (k, k', k'', k''') for which in addition the coefficient $c_{kk'k''k'''}$ in (3.34) does not vanish i.e.

$$k + k' \not\equiv 0 \mod N \quad \text{and} \quad k + k' + k'' + k''' \equiv 0 \mod N.$$

There are two quadruples (k, k', k'', k''') in K_4^N which satisfy these additional conditions,

$$\begin{cases} k + k'' = 0 \\ k' + k''' = 0 \end{cases} \quad \text{or} \quad \begin{cases} k + k''' = 0 \\ k' + k'' = 0 \end{cases}. \quad (3.39)$$

In both cases, we have $s_{k''} + s_{k'''} = -(s_k + s_{k'})$, and therefore the formula (3.33) for $c_{kk'k''k'''}$ reduces to

$$c_{kk'k''k'''} = \frac{-2|s_{k+k'}|}{|s_{k+k'}| + s_k + s_{k'}}. \quad (3.40)$$

Note that (3.40) remains valid for $k + k' = N$, since in this case $s_{k+k'} = 0$ and $s_k + s_{k'} > 0$, as k and k' must satisfy $1 \leq k, k' \leq N - 1$, but not for $k + k' = 0$, since in this case $|s_{k+k'}| + s_k + s_{k'} = 0$.

We first compute the diagonal part of $\pi_N(\frac{1}{2}\{G_\varepsilon, F_3\})$. In this case, the two possibilities in (3.39) coincide, and the solutions are

$$(k, k', k'', k''') = \begin{cases} (l,\ l,\ -l,\ -l) \\ (-l, -l,\ l,\ l) \end{cases}, \quad (3.41)$$

for some $1 \leq l \leq N - 1$. The sum of the coefficients $c_{kk'k''k'''}$ for the two cases listed in (3.41) is

$$c_{l,l,-l,-l} + c_{-l,-l,l,l} = -2|s_{2l}|\left(\frac{1}{|s_{2l}| + 2s_l} + \frac{1}{|s_{2l}| - 2s_l}\right) = \frac{-4s_{2l}^2}{s_{2l}^2 - 4s_l^2} = 4\cot^2\frac{l\pi}{N}.$$

We now turn to the off-diagonal part of $\pi_N(\frac{1}{2}\{G_3, F_3\})$. The quadruples $(k, k', k'', k''') \in K_4$ satisfying (3.39) for given $\{l, m\} \subseteq \{1, \ldots, N-1\}$ with $l < m$, $(k, k') = (\pm l, \pm m)$, and $(k'', k''') = -(k, k')$, are

$$(k, k', k'', k''') = \begin{cases} (l, & m, & -l, & -m) \\ (l, & -m, & -l, & m) \\ (-l, & m, & l, & -m) \\ (-l, & -m, & l, & m) \end{cases}. \tag{3.42}$$

The remaining matches can be obtained from (3.42) by permuting the first and second or the third and fourth columns on the right hand side of (3.42), bringing the total number of all matches to $4 \cdot 4 = 16$. Note that by formula (3.40), these permutations leave the value of the coefficients $c_{kk'k''k'''}$ invariant. Taking the sum of the coefficients $c_{kk'k''k'''}$ for all the quadruples listed in (3.42) and their permutations just mentioned, we get

$$4(c_{l,m,-l,-m} + c_{l,-m,-l,m} + c_{-l,m,l,-m} + c_{-l,-m,l,m})$$

$$= -8\left(\frac{|s_{l+m}|}{|s_{l+m}| + s_l + s_m} + \frac{|s_{l-m}|}{|s_{l-m}| + s_l - s_m}\right.$$

$$\left. + \frac{|s_{l-m}|}{|s_{l-m}| - s_l + s_m} + \frac{|s_{l+m}|}{|s_{l+m}| - s_l - s_m}\right)$$

$$= -16\left(\frac{s_{l-m}^2}{s_{l-m}^2 - (s_l - s_m)^2} + \frac{s_{l+m}^2}{s_{l+m}^2 - (s_l + s_m)^2}\right)$$

$$= \frac{-16(2 s_{l-m}^2 s_{l+m}^2 - s_{l-m}^2 (s_l + s_m)^2 - s_{l+m}^2 (s_l - s_m)^2)}{s_{l-m}^2 s_{l+m}^2 + (s_l - s_m)^2 (s_l + s_m)^2 - s_{l-m}^2 (s_l + s_m)^2 - s_{l+m}^2 (s_l - s_m)^2}$$

$$= -16,$$

since $s_{l-m}^2 s_{l+m}^2 = (s_l - s_m)^2 (s_l + s_m)^2$. Collecting the terms from the diagonal and off-diagonal parts of $\pi_N(\frac{1}{2}\{G_3, F_3\})$, we thus have

$$\pi_N\left(\frac{\alpha^2}{2}\{G_3, F_3\}\right) = \frac{\alpha^2}{16N}\left(\sum_{l=1}^{N-1} 4\cos^2\frac{\pi l}{N}|\zeta_l|^4 - 16\sum_{1 \le l < m \le N-1} s_l s_m |\zeta_l|^2 |\zeta_m|^2\right)$$

$$= \frac{\alpha^2}{4N}\left(\sum_{l=1}^{N-1}(1 - s_l^2)|\zeta_l|^4 - 2\sum_{1 \le l \ne m \le N-1} s_l s_m |\zeta_l|^2 |\zeta_m|^2\right). \tag{3.43}$$

Adding up (3.38) and (3.43), we obtain the claimed formula (3.36). □

Now we want to remove as much as possible of the term $(\mathrm{Id} - \pi_N)(\beta G_4 + \frac{\alpha^2}{2}\{G_3, F_3\})$ (the non-normal form part of $\beta G_4 + \frac{\alpha^2}{2}\{G_3, F_3\}$) from the transformed Hamiltonian (3.25), $\tilde{H}_V \circ \Psi_1$, by another coordinate transformation Ψ_2. In view of formulas (3.19) and (3.34) for G_4 and $\frac{1}{2}\{G_3, F_3\}$, respectively, and

3.1. PERIODIC CHAINS

in complete analogy to the first step we choose the transformation Ψ_2 to be of the form $\Psi_2 = X_{F_4}^t|_{t=1}$ with

$$F_4 = \sum_{(k,k',k'',k''') \in K_4 \setminus K_4^N} F^{(4)}_{kk'k''k'''} \zeta_k \zeta_{k'} \zeta_{k''} \zeta_{k'''}, \qquad (3.44)$$

where $F^{(4)}_{\sigma(k,k',k'',k''')} = F^{(4)}_{(k,k',k'',k''')}$ for any permutation $\sigma(k,k',k'',k''')$ of the quadruple $(k,k',k'',k''') \in K_4 \setminus K_4^N$. We would like to determine the coefficients of F_4 in such a way that the homological equation

$$\{G_2, F_4\} = -(\mathrm{Id} - \pi_N)\left(\beta G_4 + \frac{\alpha^2}{2}\{G_3, F_3\}\right) \qquad (3.45)$$

is satisfied. As in (3.27) one gets

$$\{G_2, F_4\} = -i \sum_{(k,k',k'',k''') \in K_4 \setminus K_4^N} (s_k + s_{k'} + s_{k''} + s_{k'''}) F^{(4)}_{kk'k''k'''} \zeta_k \zeta_{k'} \zeta_{k''} \zeta_{k'''}, \quad (3.46)$$

and (3.45) combined with (3.35) leads to the equation

$$i(s_k + s_{k'} + s_{k''} + s_{k'''}) F^{(4)}_{kk'k''k'''} \qquad (3.47)$$
$$= \frac{1}{24N}(-1)^{\varepsilon_{kk'k''k'''}}\left(\beta + \frac{3\alpha^2}{2} c^S_{kk'k''k'''}\right) \cdot \lambda_k \lambda_{k'} \lambda_{k''} \lambda_{k'''}$$

for any quadruple (k,k',k'',k''') in $K_4 \setminus K_4^N$. Here $c^S_{kk'k''k'''}$ denotes the symmetrized version of $c_{kk'k''k'''}$,

$$c^S_{kk'k''k'''} := \frac{1}{4!} \sum_{\sigma \in S_4} c_{\sigma(k,k',k'',k''')}. \qquad (3.48)$$

The following lemma due to Beukers and Rink (cf. [82]) determines the quadruples $(k,k',k'',k''') \in K_4 \setminus K_4^N$ for which $s_k + s_{k'} + s_{k''} + s_{k'''} = 0$ and therefore making (3.47) impossible to solve for all values of α and β for which $\beta + \frac{3\alpha^2}{2} c^S_{kk'k''k'''} \neq 0$. Let us introduce

$$K_4^{res} := K_{res}^+ \cup K_{res}^- \subseteq K_4$$

where

$$K_{res}^\pm := \left\{(k,k',k'',k''') \in K_4 \mid \exists l \in \mathbb{N} : 1 \leq l \leq \frac{N}{4} \text{ so that} \right.$$
$$\left. \{k,k',k'',k'''\} = \{\pm l, \pm l \mp N, \frac{N}{2} \mp l, -\frac{N}{2} \mp l\}\right\}.$$

Note that if N is odd, then $K_4^{res} = \emptyset$.

Lemma 3.1.4. *Let $(k,k',k'',k''') \in K_4 \setminus K_4^N$. Then*

$$s_k + s_{k'} + s_{k''} + s_{k'''} = 0 \text{ if and only if } (k,k',k'',k''') \in K_4^{res}.$$

In particular, if N is odd, then $s_k + s_{k'} + s_{k''} + s_{k'''} \neq 0$ for any $(k,k',k'',k''') \in K_4 \setminus K_4^N$.

For completeness, a detailed proof of Lemma 3.1.4 is given in Appendix A. As in the case of Lemma 3.1.2, it is likely that Lemma 3.1.4 also can be proved using the integrability of the Toda lattice (cf. [43]), which would be remarkable insofar as it would be a proof of an algebraic fact with methods from the theory of dynamical systems.

By Lemma 3.1.4, if N is odd, (3.47) can be solved for any $(k, k', k'', k''') \in K_4 \setminus K_4^N$, thereby determining the coefficients $F^{(4)}_{kk'k''k'''}$ with $(k, k', k'', k''') \in K_4 \setminus K_4^N$ in such a way that $F^{(4)}_{\sigma(k,k',k'',k''')} = F^{(4)}_{(k,k',k'',k''')}$ for any permutation $\sigma(k, k', k'', k''')$ of $(k, k', k'', k''') \in K_4 \setminus K_4^N$. With this choice of F_4 the canonical transformation Ψ_2 is then defined by $X_{F_4}^t|_{t=1}$. Composing Ψ_1 and Ψ_2, we obtain the transformation $\Xi := \Psi_1 \circ \Psi_2$, and we have proved the following

Proposition 3.1.5. *Assume that $N \geq 3$ is odd. The real analytic symplectic coordinate transformation $\zeta = \Xi(z)$, defined in a neighborhood of the origin in $\mathcal{Z}$, transforms the Hamiltonian $\tilde{H}_V$, given by (3.22), into its Birkhoff normal form up to order four. More precisely,*

$$\tilde{H}_V \circ \Xi = G_2 + \pi_N \left(\beta G_4 + \frac{\alpha^2}{2} \{G_3, F_3\} \right) + O(z^5), \tag{3.49}$$

with G_2 and $\pi_N(\beta G_4 + \frac{\alpha^2}{2}\{G_3, F_3\})$ given by (3.17) and (3.36), respectively.

With Proposition 3.1.5 we can now easily prove Theorem 1.2.1.

Proof of Theorem 1.2.1. Formula (3.49) provides the Taylor series expansion of $\tilde{H}_V$ in terms of the actions $I = (I_k)_{1 \leq k \leq N-1}$ given by (1.7). More precisely, $\tilde{H}_V \circ \Xi = H_{\alpha,\beta}(I) + O(z^5)$, where $H_{\alpha,\beta}(I)$ is given by

$$2 \sum_{k=1}^{N-1} s_k I_k + \frac{1}{4N} \sum_{k=1}^{N-1} (\alpha^2 + (\beta - \alpha^2)s_k^2) I_k^2 + \frac{\beta - \alpha^2}{2N} \sum_{\substack{l \neq m \\ 1 \leq l, m \leq N-1}} s_l s_m I_l I_m. \tag{3.50}$$

This completes the proof of Theorem 1.2.1. □

Now we assume that N is even, and our goal is to obtain the normal form of the FPU Hamiltonian as claimed in Theorem 1.2.4. According to Lemma 3.1.4, equation (3.47) might have no solution $F^{(4)}_{kk'k''k'''}$ for $(k, k', k'', k''') \in K_4^{res}$. We first compute the projection $\pi_{res}(\beta G_4 + \frac{\alpha^2}{2}\{G_3, F_3\})$ of the fourth-order term $\beta G_4 + \frac{\alpha^2}{2}\{G_3, F_3\}$ onto those terms which are indexed by quadruples $(k, k', k'', k''') \in K_4^{res}$, i.e. the projection onto the resonant non-normal form part of $\beta G_4 + \frac{\alpha^2}{2}\{G_3, F_3\}$.

Lemma 3.1.6. *Assume that N is even. The resonant non-normal form part of $\beta G_4 + \frac{\alpha^2}{2}\{G_3, F_3\}$ is given by*

$$\pi_{res}\left(\beta G_4 + \frac{\alpha^2}{2}\{G_3, F_3\} \right) = -\frac{\beta - \alpha^2}{4N} \left(R + R_{\frac{N}{4}} \right), \tag{3.51}$$

3.1. PERIODIC CHAINS

where

$$R = 2 \sum_{1 \leq l < \frac{N}{4}} s_{2l} \left(\zeta_l \zeta_{-N+l} \zeta_{\frac{N}{2}-l} \zeta_{-\frac{N}{2}-l} + \zeta_{-l} \zeta_{N-l} \zeta_{\frac{N}{2}+l} \zeta_{-\frac{N}{2}+l} \right) \qquad (3.52)$$

and

$$R_{\frac{N}{4}} = \begin{cases} \frac{1}{2}\left(\zeta_{\frac{N}{4}}^2 \zeta_{-\frac{3N}{4}}^2 + \zeta_{\frac{3N}{4}}^2 \zeta_{-\frac{N}{4}}^2\right) & \text{if } \frac{N}{4} \in \mathbb{N}, \\ 0 & \text{otherwise.} \end{cases} \qquad (3.53)$$

Proof. We start with the formula (3.35) for $\beta G_4 + \frac{\alpha^2}{2}\{G_3, F_3\}$. At this point we need to consider the symmetrized version (3.48) of the coefficients $c_{lml'm'}$ defined by (3.33). We claim that for any $(k_1, k_2, k_3, k_4) \in K_4^{res}$,

$$c^S_{k_1 k_2 k_3 k_4} = \frac{1}{4!} \sum_{\sigma \in S_4} c_{k_{\sigma(1)} k_{\sigma(2)} k_{\sigma(3)} k_{c(4)}} = -\frac{2}{3}. \qquad (3.54)$$

Observe that $c_{k_1 k_2 k_3 k_4}$ is invariant under the transpositions $k_1 \leftrightarrow k_2$ and $k_3 \leftrightarrow k_4$. Hence to prove (3.54) it suffices to show that

$$4(c_{k_1 k_2 k_3 k_4} + c_{k_1 k_3 k_2 k_4} - c_{k_1 k_4 k_2 k_3} + c_{k_2 k_4 k_1 k_3} + c_{k_2 k_3 k_1 k_4} + c_{k_3 k_4 k_1 k_2}) = -16. \qquad (3.55)$$

Note that any element $(k_1, k_2, k_3, k_4) \in K_4^{res}$ is, mod $2N$, a permutation of an element of the form $(l, -N+l, \frac{N}{2}-l, -\frac{N}{2}-l)$ with $1 \leq |l| \leq N/4$. For such quadruples one gets by a straightforward computation

$$c_{k_1 k_2 k_3 k_4} + c_{k_3 k_4 k_1 k_2} = -2 - 2 = -4 \qquad (3.56)$$

and, with $c_l = \cos \frac{l\pi}{N}$ (recall that $s_l = \sin \frac{l\pi}{N}$),

$$c_{k_1 k_3 k_2 k_4} + c_{k_2 k_4 k_1 k_3} = -\frac{4}{2 + 2(s_l + c_l)} - \frac{4}{2 - 2(s_l + c_l)} = -\frac{4}{s_{2l}}, \qquad (3.57)$$

and finally

$$c_{k_1 k_4 k_2 k_3} + c_{k_2 k_3 k_1 k_4} = -\frac{4}{2 + 2(s_l - c_l)} - \frac{4}{2 - 2(s_l - c_l)} = \frac{4}{s_{2l}}. \qquad (3.58)$$

Adding up (3.56)-(3.58) leads to the claimed identity (3.55).

Moreover, by the definition (3.32) of $\varepsilon_{lml'm'}$ one has that $\varepsilon_{\sigma(k_1, k_2, k_3, k_4)} = \pm 1$ for any $(k_1, k_2, k_3, k_4) \in K_4^{res}$ and any $\sigma \in S_4$, and hence

$$(-1)^{\varepsilon_{\sigma(k_1, k_2, k_3, k_4)}} = -1.$$

Further,

$$\lambda_{k_1} \lambda_{k_2} \lambda_{k_3} \lambda_{k_4} = \left|\sin \frac{l\pi}{N} \cos \frac{l\pi}{N}\right| = \frac{1}{2}\left|\sin \frac{2l\pi}{N}\right| = \frac{1}{2}|s_{2l}|.$$

Combining all these computations we get

$$\pi_{res}\left(\beta G_4 + \frac{\alpha^2}{2}\{G_3, F_3\}\right)$$

$$= \frac{1}{24N} \sum_{(k_1,k_2,k_3,k_4)\in K_4^{res}} (-1)(\beta + \frac{3\alpha^2}{2} c^S_{k_1 k_2 k_3 k_4})\lambda_{k_1}\lambda_{k_2}\lambda_{k_3}\lambda_{k_4}\zeta_{k_1}\zeta_{k_2}\zeta_{k_3}\zeta_{k_4}$$

$$= -\frac{4!\,(\beta - \alpha^2)}{24N} \sum_{1\leq l < \frac{N}{4}} \frac{s_{2l}}{2} \left(\zeta_l \zeta_{-N+l}\zeta_{\frac{N}{2}-l}\zeta_{-\frac{N}{2}-l} + \zeta_{-l}\zeta_{N-l}\zeta_{\frac{N}{2}+l}\zeta_{-\frac{N}{2}+l}\right)$$

$$\underbrace{-\frac{3!\,(\beta - \alpha^2)}{24N} \cdot \frac{1}{2}\left(\zeta^2_{\frac{N}{4}}\zeta^2_{-\frac{3N}{4}} + \zeta^2_{-\frac{N}{4}}\zeta^2_{\frac{3N}{4}}\right)}_{\text{only if } \frac{N}{4}\in\mathbb{N}} \quad (3.59)$$

$$= -\frac{\beta - \alpha^2}{4N}(R + R_{\frac{N}{4}}),$$

with R and $R_{\frac{N}{4}}$ as defined by (3.52) and (3.53), respectively. Hence Lemma 3.1.6 is proved. □

By Lemma 3.1.6, if N is even, equation (3.47) can be solved for any quadruple $(k, k', k'', k''') \in K_4 \setminus (K_4^N \cup K_4^{res})$ in such a way that $F^{(4)}_{\sigma(k,k',k'',k''')} = F^{(4)}_{(k,k',k'',k''')}$ for any permutation $\sigma(k, k', k'', k''')$ of (k, k', k'', k'''). With this choice of F_4 we then define the canonical transformation Ψ_2 by $X^t_{F_4}|_{t=1}$. Composing Ψ_1 and Ψ_2, we obtain the transformation $\Xi := \Psi_1 \circ \Psi_2$ and have proved the following analogue to Proposition 3.1.5:

Proposition 3.1.7. *Assume that N is even. The real analytic symplectic coordinate transformation $\zeta = \Xi(z)$, defined locally in a neighborhood of the origin $z = 0$ in $\mathcal{Z}$, transforms the Hamiltonian $\tilde{H}_V$, given by (3.22), into the resonant normal form up to order four,*

$$\tilde{H}_V \circ \Xi = G_2 + \pi_N\left(\beta G_4 + \frac{\alpha^2}{2}\{G_3, F_3\}\right) + \pi_{res}\left(\beta G_4 + \frac{\alpha^2}{2}\{G_3, F_3\}\right) + O(z^5),$$

with G_2, $\pi_N(\beta G_4 + \frac{\alpha^2}{2}\{G_3, F_3\})$, and $\pi_{res}(\beta G_4 + \frac{\alpha^2}{2}\{G_3, F_3\})$ given by (3.17), (3.36), and (3.51), respectively.

With Proposition 3.1.7 we can now prove Theorem 1.2.4, our normal form result on even periodic chains.

Proof of Theorem 1.2.4. We start with the formula for $\tilde{H}_V \circ \Xi$ given by Proposition 3.1.7 and treat the normal form terms $G_2 + \pi_N(\beta G_4 + \frac{\alpha^2}{2}\{G_3, F_3\})$ and the resonant normal form terms $\pi_{res}(\beta G_4 + \frac{\alpha^2}{2}\{G_3, F_3\})$ separately. With the action variables $I = (I_k)_{1\leq k\leq N-1}$ defined by (1.7) we see that $G_2 + \pi_N(\beta G_4 + \frac{\alpha^2}{2}\{G_3, F_3\}) = H_{\alpha,\beta}(I)$, where $H_{\alpha,\beta}(I)$ is defined by (3.50). Concerning the

3.1. PERIODIC CHAINS

term $\pi_{res}(\beta G_4 + \frac{\alpha^2}{2}\{G_3, F_3\})$, we first express it in terms of the real variables $(x_k, y_k)_{1 \leq k \leq N-1}$ by (3.5) related to the ζ_k's by $x_k = (\zeta_k + \zeta_{-k})/2$ and $y_k = (\zeta_{-k} - \zeta_k)/2i$. Note that

$$\zeta_l \zeta_{-N+l} \zeta_{\frac{N}{2}-l} \zeta_{-\frac{N}{2}-l} + \zeta_{-l} \zeta_{N-l} \zeta_{-\frac{N}{2}+l} \zeta_{\frac{N}{2}+l}$$
$$= 2 \operatorname{Re}(\zeta_l \zeta_{-N+l} \zeta_{\frac{N}{2}-l} \zeta_{-\frac{N}{2}-l})$$
$$= \frac{1}{2}((x_l x_{N-l} + y_l y_{N-l})(x_{\frac{N}{2}-l} x_{\frac{N}{2}+l} + y_{\frac{N}{2}-l} y_{\frac{N}{2}-l})$$
$$\quad - (x_l y_{N-l} - x_{N-l} y_l)(x_{\frac{N}{2}-l} y_{\frac{N}{2}+l} - x_{\frac{N}{2}+l} y_{\frac{N}{2}-l}))$$
$$= 2\left(J_l J_{\frac{N}{2}-l} - M_l M_{\frac{N}{2}-l}\right), \tag{3.60}$$

where for any $1 \leq k \leq N-1$, J_k and M_k are given by (1.10). Hence R, given by (3.52), can be expressed in terms of the J_k's and M_k's by

$$R(J, M) = 2 \sum_{1 \leq l < \frac{N}{4}} s_{2l} \left(\zeta_l \zeta_{-N+l} \zeta_{\frac{N}{2}-l} \zeta_{-\frac{N}{2}-l} + \zeta_{-l} \zeta_{N-l} \zeta_{-\frac{N}{2}+l} \zeta_{\frac{N}{2}+l} \right)$$
$$= 4 \sum_{1 \leq l < \frac{N}{4}} \sin \frac{2l\pi}{N} \left(J_l J_{\frac{N}{2}-l} - M_l M_{\frac{N}{2}-l} \right). \tag{3.61}$$

Similarly, if $\frac{N}{4} \in \mathbb{N}$, one concludes from (3.60) that $R_{\frac{N}{4}}$, given by (3.53), can be expressed as

$$R_{\frac{N}{4}}(J, M) = \frac{1}{2}\left(\zeta_{\frac{N}{4}}^2 \zeta_{-\frac{3N}{4}}^2 + \zeta_{\frac{3N}{4}}^2 \zeta_{-\frac{N}{4}}^2 \right) = J_{\frac{N}{4}}^2 - M_{\frac{N}{4}}^2. \tag{3.62}$$

Theorem 1.2.4 now follows from the formulas (3.50), (3.61), and (3.62). □

We now turn to the proof of Theorem 1.2.5, i.e. the integrability of the truncated Hamiltonian (1.16). Denote by $\{\cdot, \cdot\}$ the standard Poisson bracket on $\mathbb{R}^{2N-2}$. In a straightforward way one computes the Poisson brackets between the variables $I, M, J, L \in \mathbb{R}^{N-1}$, given by (1.7) and (1.10) (cf. [16], p. 28):

Lemma 3.1.8. *For any $1 \leq k, l \leq N-1$, the Poisson brackets between the variables I_k, J_k, M_k are given by*

$$\{I_l, I_k\} = \{J_l, J_k\} = \{M_l, M_k\} = 0, \tag{3.63}$$
$$\{J_l, I_k\} = -M_l(\delta_{kl} - \delta_{k+l,N}), \tag{3.64}$$
$$\{M_l, I_k\} = J_l(\delta_{kl} - \delta_{k+l,N}), \tag{3.65}$$

As a consequence, one obtains the following relations between the variables M_k, J_k, and L_k:

$$\{M_k, J_l\} = L_l(\delta_{k+l,N} - \delta_k),$$
$$\{J_k, L_l\} = M_k(\delta_{k+l,N} - \delta_{kl}),$$
$$\{L_k, M_l\} = J_l(\delta_{k+l,N} - \delta_{kl}).$$

First note that the list of functions of Theorem 1.2.5,

$$(I_k + I_{N-k})_{1 \le k \le \frac{N}{2}}, \ (I_k + I_{\frac{N}{2}+k})_{1 \le k < \frac{N}{4}}, \ (K_k)_{1 \le k \le \frac{N}{4}}, \quad (3.66)$$

contains $N-1$ terms regardless whether $\frac{N}{4}$ is an integer or not. In addition, for any $1 \le k < \frac{N}{4}$, the terms $I_k + I_{N-k}, I_{N/2-k} + I_{N/2+k}, I_k + I_{N/2+k}, K_k$ are functions of the eight variables $x_k, y_k, x_{N/2-k}, y_{N/2-k}, x_{N/2+k}, y_{N/2+k}, x_{N-k}, y_{N-k}$, the term $I_{N/2}$ is a function of the two variables $x_{N/2}, y_{N/2}$, and, in the case $\frac{N}{4} \in \mathbb{N}$, the terms $I_{\frac{N}{4}} + I_{\frac{3N}{4}}, K_{\frac{N}{4}}$ are functions of the four variables $x_{N/4}, y_{N/4}, x_{3N/4}, y_{3N/4}$. Hence we obtain a partition of the $2N - 2$ variables $x_1, y_1, \ldots, x_{N-1}, y_{N-1}$ into $\lfloor \frac{N}{4} \rfloor + 1$ pairwise disjoint sets of variables, and all Poisson brackets between variables of different sets of this partition vanish.

Lemma 3.1.9. *The $N-1$ functions listed in (3.66) are pairwise in involution.*

Proof. In view of the preceding remarks, we first rewrite the list (3.66) as

$$\left(I_k + I_{N-k}, I_{\frac{N}{2}-k} + I_{\frac{N}{2}+k}, I_k + I_{\frac{N}{2}+k}, K_k\right)_{1 \le k < \frac{N}{4}}, \quad (3.67)$$

$$I_{\frac{N}{2}}, \quad (3.68)$$

and, if $\frac{N}{4} \in \mathbb{N}$,

$$I_{\frac{N}{4}} + I_{\frac{3N}{4}}, K_{\frac{N}{4}}. \quad (3.69)$$

For any $1 \le k < \frac{N}{4}$, the four functions listed in (3.67), as well as for $k = \frac{N}{2}$ and $k = \frac{N}{4}$ (if $\frac{N}{4} \in \mathbb{N}$), the functions listed in (3.68) and (3.69), respectively, depend on only one of the $\lfloor \frac{N}{4} \rfloor + 1$ pairwise disjoint sets of variables described above. As the Poisson brackets between terms depending on variables of different sets vanish, it remains to check that functions of (3.66) with the same k are in involution with each other. In view of the formulas (1.19) and (1.21) for K_l and $K_{N/4}$, respectively (recall from (1.11) that $J_j^2 + M_j^2 = I_j I_{N-j}$ for any $1 \le j < \frac{N}{2}$), and taking into account that $(I_k)_{1 \le k \le N-1}$ are pairwise in involution, this amounts to proving that for any $1 \le l < \frac{N}{4}$,

$$\{J_l J_{\frac{N}{2}-l} - M_l M_{\frac{N}{2}-l}, I_l + I_{N-l}\} = 0, \quad (3.70)$$

$$\{J_l J_{\frac{N}{2}-l} - M_l M_{\frac{N}{2}-l}, I_{\frac{N}{2}-l} + I_{\frac{N}{2}+l}\} = 0, \quad (3.71)$$

$$\{J_l J_{\frac{N}{2}-l} - M_l M_{\frac{N}{2}-l}, I_l + I_{\frac{N}{2}+l}\} = 0, \quad (3.72)$$

and

$$\{J_{\frac{N}{4}}^2 - M_{\frac{N}{4}}^2, I_{\frac{N}{4}} + I_{\frac{3N}{4}}\} = 0. \quad (3.73)$$

First we note that by (3.64) and (3.65) one has for any $1 \le l < \frac{N}{2}$

$$\{J_l J_{\frac{N}{2}-l} - M_l M_{\frac{N}{2}-l}, I_l\} = -J_{\frac{N}{2}-l} M_l - M_{\frac{N}{2}-l} J_l \quad (3.74)$$

and

$$\{J_l J_{\frac{N}{2}-l} - M_l M_{\frac{N}{2}-l}, I_{N-l}\} = J_{\frac{N}{2}-l} M_l + M_{\frac{N}{2}-l} J_l. \quad (3.75)$$

3.2. DIRICHLET CHAINS

Since the right hand sides of (3.74) and (3.75) are invariant under exchanging l and $\frac{N}{2} - l$, the same must hold for the left hand sides, and we conclude that

$$\{J_l J_{\frac{N}{2}-l} - M_l M_{\frac{N}{2}-l}, I_{\frac{N}{2}-l}\} = -J_{\frac{N}{2}-l} M_l - M_{\frac{N}{2}-l} J_l \qquad (3.76)$$

and

$$\{J_l J_{\frac{N}{2}-l} - M_l M_{\frac{N}{2}-l}, I_{\frac{N}{2}+l}\} = J_{\frac{N}{2}-l} M_l + M_{\frac{N}{2}-l} J_l. \qquad (3.77)$$

The identities (3.70)-(3.72) now follow from the appropriate combinations of (3.74)-(3.77). In the same fashion, one concludes that (3.73) holds. □

Having shown that the functions listed in (3.67) are pairwise in involution, we can now complete the proof of Theorem 1.2.5.

Proof of Theorem 1.2.5. In view of Lemma 3.1.9, it remains to check that the quantities listed in (3.66) are functionally independent integrals. The independence is easy to verify and the fact that they are conserved quantities follows from the formula (1.18), showing that H_V^{trunc} can be written as a function of them. □

3.2 Dirichlet Chains

In this section we consider FPU chains with N' moving particles and fixed endpoints, i.e. given by the Hamiltonian (1.23) with boundary conditions (1.3), $q_0 = q_{N'+1} = 0$.

It has been observed by Rink [85] that such a chain can be treated as an invariant subsystem of a periodic lattice with $N = 2N' + 2$ particles: Let $T^*\mathbb{R}^N$ be endowed with the canonical symplectic structure and consider the linear map $S : T^*\mathbb{R}^N \to T^*\mathbb{R}^N$, defined by

$$((q_i)_{1 \leq i \leq N}, (p_i)_{1 \leq i \leq N}) \mapsto (-(q_{N-1}, \ldots, q_1, q_N), -(p_{N-1}, \ldots, p_1, p_N)). \qquad (3.78)$$

The map S is a canonical linear involution satisfying $H_V \circ S = H_V$, and we denote by $\text{Fix}(S)$ the fixed point set of S. In other words, $\text{Fix}(S)$ is the subset of all elements (q, p) in $T^*\mathbb{R}^N$ satisfying

$$(q_n, p_n) = -(q_{N-n}, p_{N-n}) \quad \forall 1 \leq n \leq N - 1 \quad \text{and} \quad q_N = p_N = 0. \qquad (3.79)$$

In particular, on $\text{Fix}(S)$, $q_N = q_{N'+1} = 0$ and $p_N = p_{N'+1} = 0$. Note that on $\text{Fix}(S)$, both the center of mass coordinate $Q = \frac{1}{N}\sum_{i=1}^N q_i$ and its momentum $P = \frac{1}{N}\sum_{i=1}^N p_i$ identically vanish. Hence $\text{Fix}(S) \subseteq \mathcal{M}$, where

$$\mathcal{M} := \{(q,p) \in T^*\mathbb{R}^N | Q = 0; P = 0\};$$

we endow $\mathcal{M}$ with the symplectic structure induced from $T^*\mathbb{R}^N$.

The phase space of an FPU chain with N' moving particles satisfying Dirichlet boundary conditions (1.3) is $T^*\mathbb{R}^{N'}$, endowed with the canonical symplectic structure $\sum_{i=1}^{N'} dq_i \wedge dp_i$. It can be embedded into $\mathcal{M}$ by the map $\Theta : T^*\mathbb{R}^{N'} \to \mathcal{M}$, given by

$$(q_i, p_i)_{1 \leq i \leq N'} \mapsto \frac{1}{\sqrt{2}}((q_i, p_i)_{1 \leq i \leq N'}, (0,0), -(q_{N'-i}, p_{N'-i})_{0 \leq i \leq N'-1}, (0,0)).$$

Note that $\Theta(T^*\mathbb{R}^{N'}) = \text{Fix}(S)$, which means that Θ is a parametrization of $\text{Fix}(S)$, and the pullback of the canonical symplectic form on $\mathcal{M}$ by Θ is the canonical symplectic structure on $T^*\mathbb{R}^{N'}$, i.e. Θ is canonical. It follows that $\text{Fix}(S)$ is a symplectic submanifold of $\mathcal{M}$.

We now express the equations (3.79) defining $\text{Fix}(S)$ locally near the origin as a subset of $\mathcal{M}$ in terms of the canonical coordinates $(x_k, y_k)_{1 \leq k \leq N-1}$ provided by Theorem 1.2.4, or more conveniently, in terms of the associated complex coordinates $(\zeta_k)_{1 \leq |k| \leq N-1}$, defined for $1 \leq k \leq N-1$ as in (3.5) by

$$\begin{cases} \zeta_k = \frac{1}{\sqrt{2}}(x_k - iy_k), \\ \zeta_{-k} = \overline{\zeta_k} = \frac{1}{\sqrt{2}}(x_k + iy_k). \end{cases} \quad (3.80)$$

Denote as in (3.6) by $\mathcal{Z}$ the linear subspace of $\mathbb{C}^{2N-2}$ consisting of such vectors $(\zeta_k)_{1 \leq |k| \leq N-1}$ satisfying (3.80). We also recall from the previous section the notations

$$c_n := \cos \frac{n\pi}{N}, \quad s_n := \sin \frac{n\pi}{N} \quad (n \in \mathbb{Z}).$$

Define the map $S_\mathcal{Z} : \mathcal{Z} \to \mathcal{Z}$ by

$$(\zeta_k)_{1 \leq |k| \leq N-1} \mapsto (-e^{4\pi i k/N} \zeta_{N-k})_{1 \leq |k| \leq N-1}. \quad (3.81)$$

Similarly to the map $S : \mathcal{M} \to \mathcal{M}$, $S_\mathcal{Z}$ is a canonical linear involution. In fact, the maps S and $S_\mathcal{Z}$ are conjugate to each other under the coordinate change of Theorem 1.2.4. Before making this statement more precise, we introduce a parametrization of the fixed point set $\text{Fix}(S_\mathcal{Z})$ of the map $S_\mathcal{Z}$. Consider the set

$$\mathcal{Z}_{Dir} := \{(\zeta_k)_{1 \leq |k| \leq N'} \in \mathbb{C}^{2N'} | \overline{\zeta}_k = \zeta_{-k} \quad \forall 1 \leq k \leq N'\},$$

endowed with the *canonical* symplectic structure induced from $\mathbb{C}^{2N'}$, and the embedding $\Theta_\mathcal{Z} : \mathcal{Z}_{Dir} \to \mathcal{Z}$ mapping $(\zeta_k)_{1 \leq |k| \leq N'}$ to the element $(\tilde{\zeta}_k)_{1 \leq |k| \leq N-1} \in \mathcal{Z}$, given by

$$\frac{1}{\sqrt{2}} \left((\zeta_k)_{1 \leq |k| \leq N'}, (0,0), (-e^{4\pi i k/N} \zeta_{N'+1-k})_{1 \leq |k| \leq N'} \right). \quad (3.82)$$

Note that $\Theta_\mathcal{Z}(\mathcal{Z}_{Dir}) = Fix(S_\mathcal{Z})$, which means that $\Theta_\mathcal{Z}$ is a parametrization of $\text{Fix}(S_\mathcal{Z})$.

Lemma 3.2.1. *In terms of the complex variables $(\zeta_k)_{1 \leq |k| \leq N-1}$ defined by Theorem 1.2.4, near the origin, the map S is given by $S_\mathcal{Z}$. More precisely, if Ψ, defined near $0 \in \mathcal{Z}$, is the coordinate transformation given by Theorem 1.2.4,*

3.2. DIRICHLET CHAINS

then $S \circ \Psi = \Psi \circ S_Z$. In particular, locally near the origin, the set $Fix(S_Z) \subseteq Z$, described by the equations

$$e^{-2\pi ik/N}\zeta_k + e^{2\pi ik/N}\zeta_{N-k} = 0 \quad (1 \leq |k| \leq N-1), \tag{3.83}$$

is the image of the set $Fix(S)$ under Ψ^{-1}. Expressed in terms of the real variables $(x_k, y_k)_{1 \leq k \leq N-1}$, the conditions (3.83) are given by

$$\begin{pmatrix} c_{2k} & -s_{2k} \\ s_{2k} & c_{2k} \end{pmatrix} \begin{pmatrix} x_k \\ y_k \end{pmatrix} + \begin{pmatrix} c_{2k} & s_{2k} \\ -s_{2k} & c_{2k} \end{pmatrix} \begin{pmatrix} x_{N-k} \\ y_{N-k} \end{pmatrix} = \begin{pmatrix} 0 \\ 0 \end{pmatrix}. \tag{3.84}$$

In particular, for $k = N' + 1 (= N/2)$ we get $\zeta_{N'+1} = 0$ and therefore

$$(x_{N'+1}, y_{N'+1}) = (0,0).$$

In order to prove Lemma 3.2.1, we need to express the map S, defined above by (3.78), with respect to the coordinates $(x_k, y_k)_{1 \leq k \leq N-1}$ of Theorem 1.2.4. The transformation, defined on a neighborhood of $0 \in Z$ in section 3.1,

$$(x, y) = (x_k, y_k)_{1 \leq k \leq N-1} \mapsto (q, p) = (q_n, p_n)_{1 \leq n \leq N} \in \mathcal{M},$$

is given by the composition $\Psi = \Psi_0 \circ \Psi_1 \circ \Psi_2$ of the canonical transformations Ψ_0, Ψ_1, and Ψ_2 introduced in section 3.1. Let $(q,p) = \Psi_0(\zeta^{(2)})$, $\zeta^{(2)} = \Psi_1(\zeta^{(3)})$, and $\zeta^{(3)} = \Psi_2(\zeta^{(4)})$, where $\zeta^{(4)} \equiv \zeta$ are the complex coordinates related to $(x_k, y_k)_{1 \leq k \leq N-1}$ by (3.80). Note that Ψ_0 is the composition of the transformations $\zeta \mapsto (v, u)$ and $(v, u) \mapsto (q, p)$, where $(v, u) = (v_i, u_i)_{1 \leq i \leq N-1}$ are the relative coordinates introduced by (3.1). Recall from section 3.1 that $(q, p) = \Psi_0(\zeta^{(2)}) \in \mathcal{M}$ is the linear transformation given by

$$-p_n + P = \frac{1}{\sqrt{N}} \sum_{1 \leq |k| \leq N-1} \sqrt{|s_k|} e^{\pi i(2n-2)k/N} \zeta_k^{(2)} \quad (1 \leq n \leq N-1), \tag{3.85}$$

$$q_{n+1} - q_n = \frac{1}{\sqrt{N}} \sum_{1 \leq |k| \leq N-1} \sqrt{|s_k|} e^{\pi i(2n-1)k/N} \zeta_k^{(2)} \quad (1 \leq n \leq N-1), \tag{3.86}$$

where $P = \frac{1}{N} \sum_{n=1}^{N} p_n$ is the total momentum of the chain. For simplicity we assume in the sequel that $P = 0$; the general case is completely analogous (in the following application to $Fix(S)$, we indeed have $P = 0$). Note that (3.85) and (3.86) continue to hold for $n = N$. For example, to see this for (3.86), we write

$$-(q_1 - q_N) = \sum_{k=1}^{N-1}(q_{k+1} - q_k)$$

and substitute the expressions (3.86) for $q_{k+1} - q_k$ into the latter sum, from which the claim follows.

Solving (3.85)-(3.86) for $(\zeta_k^{(2)})_{1 \leq |k| \leq N-1}$ one gets for $\zeta_k^{(2)} \equiv (\Psi_0^{-1}(q,p))_k$

$$\zeta_k^{(2)} = \frac{1}{2\sqrt{N|s_k|}} \sum_{n=1}^{N} \left(-e^{-\pi i(2n-2)k/N} p_n + e^{-\pi i(2n-1)k/N}(q_{n+1} - q_n) \right)$$

$$= \frac{1}{2\sqrt{N|s_k|}} \sum_{n=1}^{N} \left(-e^{-\pi i(2n-2)k/N} p_n + \left(e^{-\pi i(2n-3)k/N} - e^{-\pi i(2n-1)k/N} \right) q_n \right)$$

$$= \frac{1}{2\sqrt{N|s_k|}} \sum_{n=1}^{N} e^{-\pi i(2n-2)k/N} \left(-p_n + \left(e^{\pi ik/N} - e^{-\pi ik/N} \right) q_n \right)$$

$$= \frac{e^{2\pi ik/N}}{2\sqrt{N|s_k|}} \sum_{n=1}^{N} e^{-2\pi ink/N}(-p_n + 2is_k q_n). \tag{3.87}$$

The transformations Ψ_1 and Ψ_2 are defined locally around the origin of $\mathcal{Z}$ and are given by $\Psi_1 = X_{F_3}^t|_{t=1}$ and $\Psi_2 = X_{F_4}^t|_{t=1}$, where the Hamiltonians F_3 and F_4 are homogeneous polynomials of order three respectively four in $(\zeta_k)_{1 \le |k| \le N-1}$. Inverting Ψ_1 and Ψ_2, we obtain

$$\zeta^{(3)} = (X_{F_3}^t|_{t=-1})(\zeta^{(2)}) \quad \text{and} \quad \zeta^{(4)} = (X_{F_4}^t|_{t=-1})(\zeta^{(3)}). \tag{3.88}$$

Proof of Lemma 3.2.1. Recall first that Fix(S), defined in terms of the coordinates (q, p) by (3.79), is a symplectic submanifold of $\mathcal{M} \subseteq T^*\mathbb{R}^N$, and that all three transformations Ψ_0, Ψ_1, and Ψ_2 are canonical. We first show that $S_\mathcal{Z} \circ \Psi_0^{-1} = \Psi_0^{-1} \circ S$, and then that $S_\mathcal{Z}$ commutes with Ψ_1 and Ψ_2 (cf. (3.81) for the definition of $S_\mathcal{Z}$). This then shows the claimed identity $S \circ \Psi = \Psi \circ S_\mathcal{Z}$.

To prove that $S_\mathcal{Z} \circ \Psi_0^{-1} = \Psi_0^{-1} \circ S$, we compute for any $(q, p) \in$ Fix(S) and any $1 \le k \le N-1$

$$(S_\mathcal{Z}(\Psi_0^{-1}(q,p)))_k = -e^{4\pi ik/N}(\Psi_0^{-1}(q,p))_{N-k}$$

$$= -e^{4\pi ik/N} \frac{e^{-2\pi ik/N}}{2\sqrt{N|s_{N-k}|}} \sum_{l=1}^{N} e^{-2\pi il(N-k)/N}(-p_l + 2is_{N-k} q_l)$$

$$= \frac{e^{2\pi ik/N}}{2\sqrt{N|s_k|}} \sum_{l=1}^{N} e^{2\pi ilk/N}(p_l - 2is_k q_l),$$

using the formulas (3.81) and (3.87) for $S_\mathcal{Z}$ and Ψ_0^{-1} in the first and second equalities, respectively, and the identity $s_{N-k} = s_k$ in the third equality. Note that the vanishing summand $l = N$ can be omitted in the latter sum (it vanishes by the equations (3.79) defining Fix(S)), in which we now substitute l by $N-l$, use the notation $S(q,p) = (\tilde{S}(q), \tilde{S}(p))$, and obtain

$$(S_\mathcal{Z}(\Psi_0^{-1}(q,p)))_k = \frac{e^{2\pi ik/N}}{2\sqrt{N|s_k|}} \sum_{l=1}^{N-1} e^{2\pi i(N-l)k/N}(p_{N-l} - 2is_k q_{N-l})$$

$$= \frac{e^{2\pi ik/N}}{2\sqrt{N|s_k|}} \sum_{l=1}^{N} e^{-2\pi ilk/N}(-\tilde{S}(p)_l + 2is_k \tilde{S}(q)_l)$$

$$= (\Psi_0^{-1}(S(q,p)))_k.$$

In the second equality, we have again included the (vanishing) summand $l = N$. In particular we have shown that Fix($S_\mathcal{Z}$) is the image of Fix(S) under Ψ_0^{-1}.

3.2. DIRICHLET CHAINS

Next we claim that $F_3 : \mathcal{Z} \to \mathbb{C}$ is invariant under $S_\mathcal{Z}$, i.e. that $F_3 \circ S_\mathcal{Z} = F_3$. Recall from (3.26) and (3.31) that F_3 is given by the polynomial

$$F_3 = \sum_{(k,k',k'') \in K_3} F^{(3)}_{kk'k''} \zeta_k \zeta_{k'} \zeta_{k''},$$

with K_3 denoting the set of all $(k, k', k'') \in \mathbb{Z}^3$ satisfying $1 \leq |k|, |k'|, |k''| \leq N-1$ and $k + k' + k'' \equiv 0 \mod N$, and

$$F^{(3)}_{kk'k''} = \frac{(-1)^{(k+k'+k'')/N}}{12i\sqrt{N}} \frac{\sqrt{|s_k s_{k'} s_{k''}|}}{s_k + s_{k'} + s_{k''}}.$$

As $s_{N-k} = \varepsilon_k$ one has

$$F^{(3)}_{N-k,N-k',N-k''} = -F^{(3)}_{kk'k''} \tag{3.89}$$

for any $(k, k', k'') \in K_3$, and $(k, k', k'') \in K_3$ if and only if $(N-k, N-k', N-k'') \in K_3$. Here we view $N-k$, $N-k'$, and $N-k''$ mod $2N$ and replace them if necessary by representatives in $\{\pm 1, \ldots, \pm(N-1)\}$. Hence

$$F_3(S_\mathcal{Z}(\zeta)) = \sum_{(k,k',k'') \in K_3} F^{(3)}_{kk'k''} (S_\mathcal{Z}(\zeta))_k (S_\mathcal{Z}(\zeta))_{k'} (S_\mathcal{Z}(\zeta))_{k''}$$

$$= \sum_{(k,k',k'') \in K_3} F^{(3)}_{kk'k''} (-1)^3 e^{-4\pi i(k+k'+k'')/N} \zeta_{N-k} \zeta_{N-k'} \zeta_{N-k''}$$

$$= -\sum_{(k,k',k'') \in K_3} F^{(3)}_{kk'k''} \zeta_{N-k} \zeta_{N-k'} \zeta_{N-k''}$$

$$= -\sum_{(k,k',k'') \in K_3} F^{(3)}_{N-k,N-k',N-k''} \zeta_k \zeta_{k'} \zeta_{k''}$$

$$= \sum_{(k,k',k'') \in K_3} F^{(3)}_{kk'k''} \zeta_k \zeta_{k'} \zeta_{k''}$$

$$= F_3(\zeta),$$

where we used (3.89) in the last step.

From $F_3 \circ S_\mathcal{Z} = F_3$ it follows (by considering a Taylor expansion of Ψ_1) that near $0 \in \mathcal{Z}$ where Ψ_1 is defined, $\Psi_1 = X^t_{F_3}|_{t=1}$ commutes with $S_\mathcal{Z}$. In particular, $\text{Fix}(S_\mathcal{Z})$ is invariant under the flow $X^t_{F_3}$.

It remains to show that near $0 \in \mathcal{Z}$ where Ψ_2 is defined, $\Psi_2 = X^t_{F_4}|_{t=1}$ commutes with $S_\mathcal{Z}$. As above, this follows from $F_4 \circ S_\mathcal{Z} = F_4$. Recall from (3.44) and (3.47) that F_4 is given by the polynomial

$$F_4 = \sum_{(k,k',k'',k''') \in K_4 \setminus K_4^N} F^{(4)}_{kk'k''k'''} \zeta_k \zeta_{k'} \zeta_{k''} \zeta_{k'''}.$$

Here $K_4 \setminus K_4^N$ denotes the set of all quadruples $(k, k', k'', k''') \in \mathbb{Z}^4$ satisfying $1 \leq |k|, |k'|, |k''|, |k'''| \leq N-1$ and $k + k' + k'' + k''' \equiv 0 \mod N$ such that there do not exist $1 \leq l \leq m \leq N-1$ with $\{k, k', k'', k'''\} = \{l, m, -l, -m\}$, and

$$F^{(4)}_{kk'k''k'''} = \frac{(-1)^{(k+k'+k''+k''')/N}}{24i\sqrt{N}} \frac{(\beta + \frac{3\alpha^2}{2}c^S_{kk'k''k'''})\sqrt{|s_k s_{k'} s_{k''} s_{k'''}|}}{s_k + s_{k'} + s_{k''} + s_{k'''}}, \quad (3.90)$$

where $c^S_{kk'k''k'''} := \frac{1}{4!}\sum_{\sigma \in S_4} c_{\sigma(k,k',k'',k''')}$ and by (3.33),

$$c_{lml'm'} = \begin{cases} \frac{1}{-1+\frac{s_{l'}+s_{m'}}{|s_{l'}+m'|}} - \frac{1}{1+\frac{s_l+s_m}{|s_{l+m}|}} & \text{if } l+m \not\equiv 0 \bmod N, \\ 0 & \text{otherwise.} \end{cases} \quad (3.91)$$

First note that $(k, k', k'', k''') \in K_4 \setminus K_4^N$ if and only if $(N-k, N-k', N-k'', N-k''') \in K_4 \setminus K_4^N$ (where here we again view $N-k, \ldots, N-k'''$ mod $2N$ and replace them if necessary by representatives in $\{\pm 1, \ldots, \pm(N-1)\}$). Next, it follows from the definition (3.91) of $c_{kk'k''k'''}$ and the identity $s_{N-k} = s_k$ that $c_{N-k,N-k',N-k'',N-k'''} = c_{kk'k''k'''}$. Hence $c^S_{N-k,N-k',N-k'',N-k'''} = c^S_{kk'k''k'''}$ and, by the definition (3.90) of $F^{(4)}_{kk'k''k'''}$,

$$F^{(4)}_{N-k,N-k',N-k'',N-k'''} = F^{(4)}_{kk'k''k'''} \quad (3.92)$$

for any $(k, k', k'', k''') \in K_4 \setminus K_4^N$. Thus, using (3.92) in the last step,

$$F_4(S_Z(\zeta)) = \sum_{(k,k',k'',k''') \in K_4 \setminus K_4^N} F^{(4)}_{kk'k''k'''}(S_Z(\zeta))_k (S_Z(\zeta))_{k'} (S_Z(\zeta))_{k''} (S_Z(\zeta))_{k'''}$$

$$= \sum_{(k,k',k'',k''') \in K_4 \setminus K_4^N} F^{(4)}_{kk'k''k'''}(-1)^4 e^{-4\pi i(k+k'+k''+k''')/N} \zeta_{N-k}\zeta_{N-k'}\zeta_{N-k''}\zeta_{N-k'''}$$

$$= \sum_{(k,k',k'',k''') \in K_4 \setminus K_4^N} F^{(4)}_{kk'k''k'''}\zeta_{N-k}\zeta_{N-k'}\zeta_{N-k''}\zeta_{N-k'''}$$

$$= \sum_{(k,k',k'',k''') \in K_4 \setminus K_4^N} F^{(4)}_{N-k,N-k',N-k'',N-k'''}\zeta_k\zeta_{k'}\zeta_{k''}\zeta_{k'''}$$

$$= F_4(\zeta).$$

This proves $F_4 \circ S_Z = F_4$. Therefore (again by considering a Taylor expansion of Ψ_2), near $0 \in Z$ where Ψ_2 is defined, $\Psi_2 = X^t_{F_4}|_{t=1}$ commutes with S_Z. $\square$

Corollary 3.2.2. *On* $Fix(S_Z)$, *for any* $1 \leq k \leq \frac{N}{2}$,

$$I_k = I_{N-k} \quad (3.93)$$

and

$$J_k J_{\frac{N}{2}-k} - M_k M_{\frac{N}{2}-k} = I_k I_{\frac{N}{2}-k}. \quad (3.94)$$

Moreover

$$I_{\frac{N}{2}} = 0. \quad (3.95)$$

3.2. DIRICHLET CHAINS

Proof. In terms of the complex variables $(\zeta_k)_{1\leq |k|\leq N-1}$, we have $I_k = \zeta_k \zeta_{-k}$ for any $1 \leq k \leq N-1$. It follows from Lemma 3.2.1, in particular formula (3.83), that on $\text{Fix}(S_{\mathcal{Z}})$,

$$I_k = \zeta_k \zeta_{-k} = (-e^{4\pi i k/N}\zeta_{N-k})(-e^{-4\pi i k/N}\zeta_{-(N-k)}) = \zeta_{N-k}\zeta_{-(N-k)} = I_{N-k},$$

proving (3.93). The identity (3.95) is a consequence of $\zeta_{\frac{N}{2}}|_{Fix(S_{\mathcal{Z}})} = 0$. To prove (3.94), we first conclude from the definition (1.10) of J_k, M_k, the definition (3.80) of the complex variables $(\zeta_k)_{1\leq |k|\leq N-1}$, and the equation (3.33) describing the set $\text{Fix}(S_{\mathcal{Z}})$, that on $\text{Fix}(S_{\mathcal{Z}})$, for any $1 \leq k \leq N-1$,

$$J_k = -c_{4k}I_k,$$
$$M_k = -s_{4k}I_k.$$

It follows that on $\text{Fix}(S_{\mathcal{Z}})$,

$$J_k J_{\frac{N}{2}-k} - M_k M_{\frac{N}{2}-k} = I_k I_{\frac{N}{2}-k}(c_{4k}c_{4(\frac{N}{2}-k)} - s_{4k}s_{4(\frac{N}{2}-k)})$$
$$= I_k I_{\frac{N}{2}-k}(c_{4k}^2 + s_{4k}^2)$$
$$= I_k I_{\frac{N}{2}-k}.$$

This completes the proof of Corollary 3.2.2. □

From the definitions (1.10), (1.14), and (1.15) of the variables I_k, J_k, M_k, and the expressions R and $R_{\frac{N}{4}}$, and from the identity (3.94) from Corollary 3.2.2, we then obtain

Corollary 3.2.3. *On $\text{Fix}(S_{\mathcal{Z}})$,*

$$R = 4 \sum_{1\leq k < \frac{N}{4}} s_{2k} I_k I_{\frac{N}{2}-k} \quad \text{and} \quad R_{\frac{N}{4}} = \begin{cases} I_{\frac{N}{4}}^2 & \text{if } \frac{N}{4} \in \mathbb{N} \\ 0 & \text{otherwise.} \end{cases}$$

It follows from Corollary 3.2.3 that on $\text{Fix}(S_{\mathcal{Z}})$, the expression (1.16), the resonant fourth order normal form of periodic FPU chains with an even number N of particles, is in Birkhoff normal form up to order four. This allows us to prove Theorem 1.3.1, our normal form result on Dirichlet chains.

Proof of Theorem 1.3.1. We use the resonant normal form (1.16) for even periodic chains as starting point, $\frac{NP^2}{2} + H_{\alpha,\beta}(I) - R_{\alpha,\beta}(J,M)$, where $H_{\alpha,\beta}(I)$ and $R_{\alpha,\beta}(J,M)$ are given by (1.8) and (1.13), respectively. Using the identity (3.93) from Corollary 3.2.2, the terms in the decomposition (1.17) of $H_{\alpha,\beta}$, when restricted to $\text{Fix}(S_{\mathcal{Z}})$, are given by

$$H^{(2)}(I) = 4\sum_{k=1}^{N'} s_k I_k, \tag{3.96}$$

$$H^{(4)}_{\alpha,\beta}(I) = \frac{1}{N}\sum_{k=1}^{N'} d_k^+ I_k^2 + \frac{4(\beta-\alpha^2)}{2N} \sum_{\substack{1\leq k,l \leq N' \\ k\neq l}} s_k s_l I_k I_l, \tag{3.97}$$

and

$$\frac{1}{2N} \sum_{k=1}^{\frac{N}{2}-1} d_k^- I_k I_{N-k} = \frac{1}{2N} \sum_{k=1}^{N'} d_k^- I_k^2. \tag{3.98}$$

On $\mathrm{Fix}(S_{\mathcal{Z}})$, we conclude from Corollary 3.2.3 that

$$\begin{aligned}
-R_{\alpha,\beta}(J,M) &= -\frac{\beta-\alpha^2}{4N}\left(R(J,M) + R_{\frac{N}{4}}(J,M)\right) \\
&= -\frac{\beta-\alpha^2}{4N}\left(4 \sum_{1\leq k < \frac{N}{4}} s_{2k} I_k I_{\frac{N}{2}-k} + \underbrace{I_{\frac{N}{4}}^2}_{\text{only if } \frac{N}{4}\in\mathbb{N}}\right).
\end{aligned} \tag{3.99}$$

We then obtain the claimed formula (1.24) by adding up (3.96)-(3.99), noting that $d_k^+ + \frac{d_k^-}{2} = \frac{1}{2}(\alpha^2 + 3(\beta-\alpha^2)s_k^2)$, and replacing I_k by its pullback $\frac{1}{2}I_k$ with respect to the parametrization $\Theta_{\mathcal{Z}}$ of $\mathrm{Fix}(S_{\mathcal{Z}})$ introduced in (3.82). $\square$

Chapter 4

Nondegeneracy and Convexity

In this chapter we prove all claims on the nondegeneracy and convexity properties of the Hessians of the fourth-order Birkhoff normal forms of the odd periodic and Dirichlet FPU chains, as stated in Theorems 1.2.3 and 1.3.3, respectively.

4.1 Periodic Chains

We start with the Hessian $Q_{\alpha,\beta}$ of $H_{\alpha,\beta}(I)$ at $I = 0$, the fourth-order Birkhoff normal form of odd periodic chains, given by (1.8). In the process of investigating of $Q_{\alpha,\beta}$ we repeatedly encounter matrices of the form $E+\mathrm{diag}(\mu_1,\ldots,\mu_{N-1})$, where the $(N-1) \times (N-1)$-matrix E is given by

$$E := \begin{pmatrix} 1 & \cdots & 1 \\ \vdots & & \vdots \\ 1 & \cdots & 1 \end{pmatrix}, \qquad (4.1)$$

and with complex numbers $(\mu_k)_{1 \leq k \leq N-1}$. The determinant of such matrices $E + \mathrm{diag}(\mu_1,\ldots,\mu_{N-1})$ can be explicitly computed.

Lemma 4.1.1. *Let $(\mu_k)_{1 \leq k \leq N-1}$ be given nonzero complex numbers. Then*

$$\det(E + \mathrm{diag}(\mu_1,\ldots,\mu_{N-1})) = \left(1 + \sum_{k=1}^{N-1} \frac{1}{\mu_k}\right) \cdot \prod_{k=1}^{N-1} \mu_k. \qquad (4.2)$$

In particular, $E + \mathrm{diag}(\mu_1,\ldots,\mu_{N-1})$ is regular if and only if $\sum_{k=1}^{N-1} \frac{1}{\mu_k} \neq -1$.

Proof. Expanding the determinant of $(E + \mathrm{diag}(\mu_1,\ldots,\mu_{N-1}))$ with respect to its rows, it follows that

$$\det(E + \mathrm{diag}(\mu_1,\ldots,\mu_{N-1})) = \prod_{k=1}^{N-1} \mu_k - \sum_{k=1}^{N-1} \prod_{\substack{1 \leq l \leq N-1 \\ l \neq k}} \mu_l.$$

This leads to formula (4.2). □

First let us treat the β-chain, i.e. the case $\alpha = 0$, $\beta \neq 0$. The following proposition is related to earlier results of Rink [82].

Proposition 4.1.2. *Let N be odd and assume that $\alpha = 0$ in (1.4). Then the following holds:*

(i) The Birkhoff normal form of H_V up to order four is given by $\frac{NP^2}{2} + H_{0,\beta}(I)$ where

$$H_{0,\beta}(I) = 2 \sum_{k=1}^{N-1} s_k I_k + \frac{\beta}{4N} \left(\sum_{k=1}^{N-1} s_k^2 I_k^2 + 2 \sum_{\substack{l \neq m \\ 1 \leq l, m \leq N-1}} s_l s_m I_l I_m \right). \quad (4.3)$$

(ii) For any $\beta \neq 0$, $H_{0,\beta}(I)$ is nondegenerate at $I = 0$.

Proof. The fourth-order Birkhoff normal form (4.3) of H_V is given by the formula (3.50) evaluated at $\alpha = 0$. To investigate the Hessian $Q_{0,\beta}$ of $H_{0,\beta}(I)$ at $I = 0$ we write

$$Q_{0,\beta} = \frac{\beta}{4N} \Delta P \Delta, \quad (4.4)$$

where

$$\Delta = \operatorname{diag}\left(\sin \frac{k\pi}{N} \right)_{1 \leq k \leq N-1} \quad (4.5)$$

and

$$P = 2 \cdot \left(E - \frac{1}{2} \operatorname{Id}_{N-1} \right)$$

for the matrix E introduced in (4.1). In view of (4.4) and (4.5), it follows that

$$\det Q_{0,\beta} = \left(\frac{\beta}{4N} \right)^{N-1} \cdot \det P \cdot \prod_{k=1}^{N-1} \sin^2 \frac{k\pi}{N},$$

where by Lemma 4.1.1,

$$\det P = 2^{N-1} \left(1 - 2(N-1) \right) (-1/2)^{N-1} = (-1)^N (2N - 3) \neq 0.$$

Hence, if $\beta \neq 0$, $\det Q_{0,\beta} \neq 0$, and it follows that $H_{0,\beta}(I)$ is nondegenerate at $I = 0$. □

Lemma 4.1.3. *If $\beta < 0$, then $Q_{0,\beta}$ has one negative eigenvalue, whereas if $\beta > 0$, then $Q_{0,\beta}$ has $N - 2$ negative eigenvalues. In particular, for any $\beta \neq 0$, $Q_{0,\beta}$ is indefinite (and $H_{0,\beta}$ is therefore not convex). Moreover, for any $\beta \neq 0$, $H_{0,\beta}$ is quasiconvex at $I = 0$ (and therefore also directionally quasiconvex at $I = 0$).*

4.1. PERIODIC CHAINS

Proof. We use the decomposition (4.4) of $Q_{0,\beta}$ to show that $Q_{0,\beta}$ can be continuously deformed to $\frac{\beta}{4N}P$: Consider for $0 \leq t \leq 1$

$$Q_{0,\beta}(t) := \frac{\beta}{4N}(t\Delta + (1-t)\,\mathrm{Id})\, P\, (t\Delta + (1-t)\,\mathrm{Id}).$$

As the diagonal matrix $t\Delta + (1-t)\,\mathrm{Id}$ is positive definite for any $0 \leq t \leq 1$ and P is regular and symmetric, $Q_{0,\beta}(t)$ is a symmetric regular $(N-1) \times (N-1)$-matrix for any $0 \leq t \leq 1$. For $t = 0$, we have $Q_{0,\beta}(0) = \frac{\beta}{4N}P$, whereas for $t = 1$, $Q_{0,\beta}(1) = Q_{0,\beta}$. Therefore, the index (i.e. the number of negative eigenvalues) of $Q_{0,\beta}$ coincides with the index of $\frac{\beta}{4N}P$. The eigenvalues of P are $\mu_1 = 2N - 3$ with multiplicity one and $\mu_2 = -1$ with multiplicity $N - 2$. The proof of the quasiconvexity property will be given on page 56 after the proof of the corresponding statement for arbitrary $\alpha \in \mathbb{R}$ (the proof of Proposition 4.1.4 does not use the quasiconvexity properties of $Q_{0,\beta}$). $\square$

We now turn to the case $\alpha \neq 0$ and arbitrary $\beta \in \mathbb{R}$.

Proposition 4.1.4. *Assume that N is odd and $\alpha \neq 0$ in (1.4). Then, for α fixed, $\det Q_{\alpha,\beta}$ is a polynomial in β of degree $N-1$ and has $N-1$ pairwise different real zeroes which we list in increasing order and denote by $\beta_k = \beta_k(\alpha)$ ($1 \leq k \leq N-1$). They satisfy $0 < \beta_1 < \alpha^2$, $2\alpha^2 < \beta_2 < \ldots < \beta_{N-1}$ and contain the $(N-1)/2$ distinct numbers*

$$\alpha^2 \cdot \left(1 + \sin^{-2}\frac{k\pi}{N}\right) \quad \left(1 \leq k \leq \frac{N-1}{2}\right).$$

When considered as functions $\beta_k = \beta_k^{(N)}(\alpha)$ of N, the zeroes β_1 and β_2 satisfy

$$\beta_1 \to \alpha^2, \quad \beta_2 \to 2\alpha^2 \qquad (N \to \infty). \tag{4.6}$$

Moreover

$$\mathrm{index}(Q_{\alpha,\beta}) = \begin{cases} 1 & \text{for } \beta < \beta_1, \\ 0 & \text{for } \beta_1 < \beta < \beta_2, \\ N-2 & \text{for } \beta > \beta_{N-1}. \end{cases} \tag{4.7}$$

Hence $H_{\alpha,\beta}$ is convex at $I = 0$ if and only if $\beta_1 < \beta < \beta_2$, in particular if $\frac{\beta}{\alpha^2} \in [1,2]$. Moreover, $H_{\alpha,\beta}$ is quasiconvex at $I = 0$ if and only if $\beta \notin [\beta_2, \beta_{N-1}]$, and directionally quasiconvex at $I = 0$ for any $\beta \in \mathbb{R}$.

Proof. Since most of the statements of Propositions 4.1.4 are also true in the case where N is even we do not assume a priori that N is odd, and we will explicitly mention when we make any assumption on the parity of N.

Keep $\alpha \in \mathbb{R} \setminus \{0\}$ fixed and consider the map $\beta \mapsto \det(Q_{\alpha,\beta})$. It follows from (3.50) that $\det(Q_{\alpha,\beta})$ is a polynomial in β of degree at most $N-1$,

$$\det(Q_{\alpha,\beta}) = \sum_{j=0}^{N-1} c_j \beta^j,$$

with $c_0 = \det(Q_{\alpha,0})$ and $c_{N-1} = \det(Q_{0,1})$. By Proposition 4.1.2, $\det(Q_{0,1}) \neq 0$, hence the degree of the polynomial $\det(Q_{\alpha,\beta})$ is precisely $N-1$. We now claim that $\det(Q_{\alpha,\beta})$ has $N-1$ real zeroes (counted with multiplicities). For $|\beta|$ large enough, $\text{index}(Q_{\alpha,\beta})$ is equal to $\text{index}(Q_{0,\beta})$. By Lemma 4.1.3, $\text{index}(Q_{0,\beta})$ is $N-2$ for any $\beta > 0$ and 1 for any $\beta < 0$. Hence there exists some $R > 0$ such that $\text{index}(Q_{\alpha,\beta}) = N-2$ for any $\beta > R$ and $\text{index}(Q_{\alpha,\beta}) = 1$ for any $\beta < -R$. In the case $\beta = \alpha^2$, the matrix Q_{α,α^2} is a positive multiple of the identity matrix (which we already noticed in our computation of the Birkhoff normal form of the periodic Toda lattice [43]), hence $\text{index}(Q_{\alpha,\alpha^2}) = 0$. It then follows that $\text{index}(Q_{\alpha,\beta})$ must change at least once in the open interval $(-\infty, \alpha^2)$ and at least $N-2$ times (counted with multiplicities) in (α^2, ∞). Since a change of $\text{index}(Q_{\alpha,\beta})$ induces a zero of $\det(Q_{\alpha,\beta})$, our consideration shows that $\beta \mapsto \det(Q_{\alpha,\beta})$ has $N-1$ real zeroes (counted with multiplicities). Further we have $\beta_1(\alpha) < \alpha^2 < \beta_2(\alpha)$.

Next we prove that $\beta_1(\alpha) > 0$, i.e. that $Q_{\alpha,\beta}$ is regular for any $\beta \leq 0$. Similarly to the decomposition (4.4) for $\alpha = 0$, we write $Q_{\alpha,\beta}$ as a product,

$$Q_{\alpha,\beta} = \frac{\alpha^2 - \beta}{4N} \Delta P_{\alpha,\beta} \Delta, \tag{4.8}$$

with Δ defined by (4.5) and $P_{\alpha,\beta}$ by

$$P_{\alpha,\beta} = -2 \left(E + \text{diag}\left(-\frac{1}{2}\left(1 + \frac{\gamma(\alpha,\beta)}{\sin^2 \frac{k\pi}{N}}\right) \right)_{1 \leq k \leq N-1} \right), \tag{4.9}$$

where E is again given by (4.1) and $\gamma(\alpha,\beta)$ by

$$\gamma(\alpha,\beta) := \frac{\alpha^2}{\alpha^2 - \beta} = \frac{1}{1 - \frac{\beta}{\alpha^2}}. \tag{4.10}$$

The quantity $\gamma \equiv \gamma(\alpha,\beta)$ will play an important role in the sequel. As $-\infty < \beta \leq 0$ it follows that $0 < \gamma \leq 1$ and $-\frac{1}{2}\left(1 + \frac{\gamma}{s_k^2}\right) < 0$ for any $1 \leq k \leq N-1$. By Lemma 4.1.1, that $P_{\alpha,\beta}$ is regular if $f(\gamma) \neq 0$, where

$$f(\gamma) := 1 - 2 \sum_{k=1}^{N-1} \left(1 + \frac{\gamma}{s_k^2}\right)^{-1} = 1 - 2 \sum_{k=1}^{N-1} \frac{s_k^2}{\gamma + s_k^2}. \tag{4.11}$$

Note that $f(\gamma)$ is increasing in $0 < \gamma \leq 1$. We now assume N to be odd and estimate $f(1)$ by

$$f(1) = 1 - 4\sum_{k=1}^{(N-1)/2} \frac{s_k^2}{1+s_k^2} < 1 - 4\frac{s_{(N-1)/2}^2}{1+s_{(N-1)/2}^2} = 1 - 4\frac{c_{1/2}^2}{1+c_{1/2}^2} = -3 + \frac{4}{1+c_{1/2}^2}.$$

As for $N \geq 3$

$$-3 + \frac{4}{1+c_{1/2}^2} < -3 + \frac{4}{1+\cos^2 \frac{\pi}{6}} = -\frac{5}{7},$$

4.1. PERIODIC CHAINS

it follows that $f(1) < 0$. Hence we have shown that $f(\gamma) < 0$ for $0 < \gamma \leq 1$, and as explained above, by Lemma 4.1.1 this implies that $P_{\alpha,\beta}$ is regular for $\beta \leq 0$. Hence we have proved that $\beta_1(\alpha) > 0$.

By (4.11) and using the assumption that N is odd, one sees that f is a rational function of $\gamma \in \mathbb{R}$ with poles of order one at $\gamma := -s_k^2$ ($1 \leq k \leq \frac{N-1}{2}$). The derivative of f can be computed to be

$$f'(\gamma) = 2 \sum_{k=1}^{N-1} \frac{1}{(s_k^2 + \gamma)^2}. \tag{4.12}$$

Hence f is strictly increasing on all connected components of its domain. Furthermore, for any of the poles $(-s_k^2)_{1 \leq k \leq \frac{N-1}{2}}$ we have the asymptotic behaviour

$$f(\gamma) \to \infty \ (\gamma \nearrow -s_k^2), \quad f(\gamma) \to -\infty \ (\gamma \searrow -s_k^2). \tag{4.13}$$

In addition, one sees that

$$f(\gamma) \nearrow 1 \ (\gamma \to \infty), \quad f(\gamma) \searrow 1 \ (\gamma \to -\infty). \tag{4.14}$$

It follows from (4.12)-(4.14) that f has precisely one zero in every of the $\frac{N-3}{2}$ bounded intervals $\left(-s_k^2, -s_{k-1}^2\right)$ ($1 \leq k \leq \frac{N-3}{2}$) and precisely one zero γ_{N-1} in the unbounded interval $\left(-s_1^2, \infty\right)$ (since it is the largest zero of f, we denote this zero by γ_{N-1} even though it corresponds to β_1). The above estimate $\gamma(1) < 0$ together with $f(\gamma) \to 1$ for $\gamma \to \infty$ shows that $\gamma_{N-1} > 1$ - we will estimate γ_{N-1} more precisely below.

Next introduce $\mu_k := -\frac{1}{2}(1 + \frac{\gamma}{s_k^2})$ and note that for any β with $\gamma = -s_{k_0}^2$ for some $1 \leq k_0 \leq \frac{N-1}{2}$ (i.e. the poles of f mentioned above) one has $\mu_{k_0} = \mu_{N-k_0} = 0$. As $k_0 \neq N - k_0$ if $1 \leq k_0 \leq \frac{N-1}{2}$ it then follows that $P_{\alpha,\beta}$ has two equal rows and is therefore singular. Note that $\gamma \equiv \gamma(\alpha,\beta) = -s_{k_0}^2$ corresponds to $\frac{\beta}{\alpha^2} = 1 + s_{k_0}^{-2}$, and that $\gamma \in \left(-s_k^2, -s_{k-1}^2\right)$ corresponds to $\frac{\beta}{\alpha^2} \in \left(1+s_k^{-2}, 1+s_{k-1}^{-2}\right) \subset [2, \infty)$ for any $2 \leq k \leq \frac{N-1}{2}$. Finally, as mentioned before, $\gamma_{N-1} > 1$ corresponds to $0 < \frac{\beta_1}{\alpha^2} < 1$. Alltogether, we have proved that the function $\beta \mapsto \det(Q_{\alpha,\beta})$ has precisely $N-1$ pairwise different zeroes on $\mathbb{R}$. The formula (4.7) on the index of $Q_{\alpha,\beta}$ easily follows from the above analysis.

Since $Q_{\alpha,\beta}$ is *not* nondegenerate for these $N-1$ zeroes $(\beta_k(\alpha))_{1 \leq k \leq N-1}$, the question naturally arises whether some weaker version of Kolmogorov's nondegeneracy condition is satisfied in these cases. Preliminary computations suggest that at least for $\gamma = -s_{k_0}^2 = -\sin^2 \frac{k_0 \pi}{N}$, the isoenergetic nondegeneracy condition (2.11) is not satisfied. However, it is still possible that Rüssmann's higher order nondegeneracy conditions are fulfilled for some or all of the exceptional $\beta's$. Alltogether, it remains an open question whether some variant of the KAM theorem can be applied to odd periodic FPU chains in these exceptional cases.

We now prove the quasiconvexity and directional quasiconvexity properties of $H_{\alpha,\beta}$, and we first turn to the claim that $H_{\alpha,\beta}(I)$ is directionally quasiconvex at $I = 0$. By the definition (2.17) of directional quasiconvexity, we have to show that $\xi = 0$ is the only solution in $\mathbb{R}_{\geq 0}^{N-1}$ of the system of equations in (2.14). Recall from (1.6) that in our case, $\omega_k = 2\sin\frac{k\pi}{N}$ for any $1 \leq k \leq N-1$. Hence for any $\xi_1, \ldots, \xi_{N-1} \geq 0$, $\langle \omega, \xi \rangle = 2\sum_{k=1}^{N-1} s_k \xi_k > 0$, unless $\xi_1 = \ldots = \xi_{N-1} = 0$. This proves the claim.

To prove the (non-directional) quasiconvexity properties, we have to investigate real solutions $\xi_1, \ldots, \xi_{N-1}$ of (2.14) without the restriction $\xi \in \mathbb{R}_{\geq 0}^{N-1}$. In greater detail, by (3.50), (2.14) reads

$$\frac{\beta - \alpha^2}{4N}\left(\sum_{k=1}^{N-1}(-\gamma + s_k^2)\xi_k^2 + 2\sum_{1\leq l\neq m\leq N-1} s_l s_m \xi_l \xi_m\right) = 0, \qquad (4.15)$$

$$2\sum_{k=1}^{N-1} s_k \xi_k = 0. \qquad (4.16)$$

Note that

$$\sum_{1\leq l\neq m\leq N-1} s_l s_m \xi_l \xi_m = \sum_{1\leq l,m\leq N-1} s_l s_m \xi_l \xi_m - \sum_{k=1}^{N-1} s_k^2 \xi_k^2 = \left(\sum_{k=1}^{N-1} s_k \xi_k\right)^2 - \sum_{k=1}^{N-1} s_k^2 \xi_k^2. \qquad (4.17)$$

Hence, substituting (4.16) into (4.15), we obtain

$$\sum_{k=1}^{N-1}(-\gamma + s_k^2)\xi_k^2 - 2\sum_{k=1}^{N-1} s_k^2 \xi_k^2 = -\sum_{k=1}^{N-1}(-\gamma + s_k^2)\xi_k^2 = 0. \qquad (4.18)$$

One now sees that (4.18) has a nontrivial solution if and only if $\gamma + s_1^2 \leq 0$ and $\gamma + s_{(N-1)/2}^2 \geq 0$, i.e. if $\gamma \in [-1 + s_{1/2}^2, -s_1^2]$. This corresponds to $\frac{\beta}{\alpha^2} \in [1 + \frac{1}{1-s_{1/2}^2}, 1 + \frac{1}{s_1^2}] = [\frac{\beta_1}{\alpha^2}, \frac{\beta_{N-1}}{\alpha^2}]$. Conversely, if $\gamma \in [-1 + s_{1/2}^2, -s_1^2]$, one easily finds a nontrivial solution of (4.18) and then of (4.15)-(4.16).

The case $\alpha = 0$ is equivalent to $\gamma = 0$, which can be explicitly checked by noting that by substituting (4.3) into the quasiconvexity condition (2.14), we obtain the system (4.15)-(4.16) for $\gamma = 0$. The above considerations then show that in this case, (4.15)-(4.16) does not have a nontrivial solution, which proves that $H_{0,\beta}$ is quasiconvex at $I = 0$ for any $\beta \in \mathbb{R}$. This completes the proof of the quasiconvexity claims for odd periodic FPU chains, and thereby also the proof of Lemma 4.1.3.

We now turn to the asymptotic statements (4.6). It follows from the above analysis that

$$\beta_2^{(N)} = \alpha^2\left(1 + \frac{1}{1 - \sin^2\frac{\pi}{2N}}\right),$$

and one sees that $\beta_2^{(N)} \to 2\alpha^2$ for $N \to \infty$. On the other hand, proving that $\beta_1^{(N)} \to \alpha^2$ for $N \to \infty$ turns out to be considerably more difficult; nevertheless, we consider it justified to prove this fact in detail since the two asymptotic

4.1. PERIODIC CHAINS

statements together give us precise information on the length of the "interval of convexity" of $Q_{\alpha,\beta}$ in the limit $N \to \infty$; moreover, we consider the combinatorial method used in the proof of Lemma 4.1.5 below to be of some interest.

We mentioned above that $\beta \to \det Q_{\alpha,\beta}$ has $N-1$ pairwise distinct zeros, of which $\frac{N-1}{2}$ ones are in terms of $\gamma \equiv \gamma(\alpha,\beta)$ given by $\gamma = -s_k^2$, $1 \le k \le \frac{N-1}{2}$, and the other $\frac{N-1}{2}$ ones are zeroes of the meromorphic function f defined in (4.11), $f(\gamma) = 1 - 2\sum_{k=1}^{N-1}\left(1 + \gamma/s_k^2\right)^{-1}$, whose $\frac{N-1}{2}$ poles of order one are exactly the other zeros of $\beta \to \det Q_{\alpha,\beta}$.

Thus, if we multiply $f(\gamma)$ by $\prod_{k=1}^{N-1}(\gamma + s_k^2) = \left(\prod_{k=1}^{(N-1)/2}(\gamma + s_k^2)\right)^2$, we obtain a polynomial p whose $N-1$ zeros are precisely the $N-1$ zeros of the function $\beta \mapsto \det Q_{\alpha,\beta}$. (The zeroes of f are also zeroes of p, and the first-order poles of f are first-order zeroes of p.) We will show that the largest zero γ_{N-1} of $p(\gamma)$ satisfies $N-1 < \gamma_{N-1} < N$, from which it will follow that $\beta_1 \to \alpha^2$ for $N \to \infty$. Explicitly, the polynomial $p(\gamma)$ is then given by

$$p(\gamma) = \left(\prod_{k=1}^{N-1}(\gamma + s_k^2)\right) \cdot \left(1 - 2\sum_{k=1}^{N-1}\frac{s_k^2}{\gamma + s_k^2}\right)$$
$$= \prod_{k=1}^{N-1}(\gamma + s_k^2) - 2\sum_{k=1}^{N-1}s_k^2 \prod_{\substack{1 \le l \le N-1 \\ l \ne k}}(\gamma + s_l^2)$$

Note that when $p(\gamma)$ is ordered by powers of γ, its coefficients are *symmetric* polynomials of the $N-1$ variables $(s_k)_{1 \le k \le N-1}$. In other words, we can express these coefficients through the N elementary symmetric polynomials $(\Pi_n(t_1,\ldots,t_{N-1}))_{0 \le n \le N-1}$ evaluated for $t_k := s_k^2$. These elementary symmetric polynomials are given by

$$\Pi_0 := 1, \tag{4.19}$$

$$\Pi_n(t_1,\ldots,t_{N-1}) := \sum_{1 \le i_1 < \ldots < i_n \le N-1} t_{i_1} \cdot \ldots \cdot t_{i_n} \quad (1 \le n \le N-1). \tag{4.20}$$

Ordering $p(\gamma)$ by powers of γ, this leads to

$$p(\gamma) = \sum_{r=0}^{N-1} \Pi_r(s_1^2,\ldots,s_{N-1}^2)(1 - 2r)\gamma^{N-1-r}. \tag{4.21}$$

We now evaluate the polynomials $(\Pi_n)_{0 \le n \le N-1}$ given by (4.19) and (4.20) for $t_k = s_k^2$. We give the (combinatorial) proof of the following lemma in Appendix B.

Lemma 4.1.5. *For any* $0 \le r \le N-1$,

$$\Pi_r\left(s_1^2,\ldots,s_{N-1}^2\right) = 4^{-r}\frac{N}{N-r}\binom{2N-r-1}{r}. \tag{4.22}$$

Substituting (4.22) into (4.21), we obtain

$$p(\gamma) = \sum_{r=0}^{N-1}(1-2r)\Pi_r(s_1^2,\ldots,s_{N-1}^2)\gamma^{N-1-r}$$

$$= 4^{-(N-1)}\sum_{r=0}^{N-1}(1-2r)\frac{N}{N-r}\binom{2N-r-1}{r}(4\gamma)^{N-1-r}.$$

The index substitution $s = N-1-r$ then leads to (note that $1-2r = 3-2(N-s)$)

$$p(\gamma) = 4^{-(N-1)}\sum_{s=0}^{N-1}(3-2(N-s))\frac{N}{s+1}\binom{N+s}{N-1-s}(4\gamma)^s.$$

We now omit the factor $4^{-(N-1)}$ (since it does not influence the zeros of p) and set $x := 4\gamma$, which leads to the polynomial

$$P(x) := \sum_{s=0}^{N-1}\frac{(3-2(N-s))N}{s+1}\binom{N+s}{N-1-s}x^s$$

$$= x^{N-1} - 2N\cdot x^{N-2} + O(x^{N-3}).$$

By Vieta's theorem (see e.g. [104]), the $N-1$ zeros $x_1,\ldots,x_{N-1}$ of P must satisfy

$$\sum_{k=1}^{N-1} x_k = 2N. \qquad (4.23)$$

Recall from above that we already know precisely $\frac{N-1}{2}$ zeroes, namely $x_k = -4\sin^2\frac{k\pi}{N}$ ($1 \leq k \leq \frac{N-1}{2}$), and from $\frac{N-3}{2}$ zeroes we know that they satisfy $-4\sin^2\frac{k\pi}{N} < x_k < -4\sin^2\frac{(k+1)\pi}{N}$ ($\frac{N+1}{2} \leq k \leq N-2$). Hence, by (4.23), the remaining zero x_{N-1} is located in the open interval

$$\left(2N + 4\sum_{k=1}^{\frac{N-1}{2}}\sin^2\frac{k\pi}{N} + 4\sum_{k=\frac{N+3}{2}}^{N-1}\sin^2\frac{k\pi}{N},\ 2N + 4\sum_{k=1}^{\frac{N-1}{2}}\sin^2\frac{k\pi}{N} + 4\sum_{k=\frac{N+1}{2}}^{N-2}\sin^2\frac{k\pi}{N}\right)$$

$$= \left(2N + 4\cdot\left(\frac{N}{2} - \sin^2\frac{(N-1)\pi}{2N}\right),\ 2N + 4\cdot\left(\frac{N}{2} - \sin^2\frac{\pi}{N}\right)\right)$$

$$= \left(4N - 4\sin^2\left(\frac{\pi}{2}\left(1 - \frac{1}{N}\right)\right),\ 4N - 4\sin^2\frac{\pi}{N}\right) \subset (4(N-1), 4N).$$

In terms of $\gamma = x/4$, we conclude that the zero γ_{N-1} satisfies

$$N - 1 < \gamma_{N-1} < N. \qquad (4.24)$$

Numerical evidence (Mathematica computations) actually suggests that we have the asymptotic formula $\gamma_{N-1} = N - \frac{3}{4} + o(1)$ ($N \to \infty$), in accordance with the

4.2. DIRICHLET CHAINS

analytically derived estimate (4.24). In terms of $\beta = \alpha^2(1 - \frac{1}{\gamma})$, it follows that the corresponding zero β_1 can be estimated by

$$\alpha^2\left(1 - \frac{1}{N-1}\right) < \beta_1 < \alpha^2\left(1 - \frac{1}{N}\right),$$

in particular, by the "sandwich theorem", we have $\beta_1 \to \alpha^2$ for $N \to \infty$, as claimed. □

Proof of Theorem 1.2.3. Part (i) is proved by Proposition 4.1.4, and (ii) follows from Proposition 4.1.2 and Lemma 4.1.3. □

We again emphasize that we do not have an analogue to Theorem 1.2.3 for even periodic chains, since in this case the fourth order normal form contains the resonant term $R_{\alpha,\beta}(J, M)$ - see Theorem 1.2.4. By Theorem 1.2.5 however, the (truncated) resonant fourth order normal form H_V^{trunc} of even periodic chains is an integrable system, and in (1.18) we have deduced an explicit formula for H_V^{trunc} in terms of the integrals. We think that it should be possible that by imitating a procedure of San [87] and Zung [105] for other types of singularities of simpler systems, it could be explicitly shown that H_V^{trunc} is nondegenerate with respect to these integrals.

4.2 Dirichlet Chains

We now turn to the chains with N' particles and Dirichlet boundary conditions and investigate the Hessian $Q_{\alpha,\beta}^D$ of $H_{\alpha,\beta}^D$ at $I = 0$, the fourth order Birkhoff normal form of Dirichlet chains, given by (1.24). Again, we first consider the β-chain, i.e. the case $\alpha = 0$, $\beta \neq 0$. Note that as before (cf. (3.7)), the numbers $(s_k)_{1 \leq k \leq N'}$ are defined by $s_k = \sin \frac{k\pi}{N}$, which however, by the treatment of N'-particle Dirichlet chains as invariant submanifolds of periodic chains with $N = 2N' + 2$ particles (explained in section 1.3), should now be read as $s_k = \sin \frac{k\pi}{2N'+2}$.

Proposition 4.2.1. *Assume that $\alpha = 0$ in (1.4). Then the following holds:*

(i) *The Birkhoff normal form of H_V^D with Dirichlet boundary conditions up to order four is given by $\frac{(N'+1)P^2}{2} + H_{0,\beta}^D(I)$ where*

$$H_{0,\beta}^D(I) = 2\sum_{k=1}^{N'} s_k I_k + \frac{\beta}{16(N'+1)}\left(\sum_{k=1}^{N'} 3s_k^2 I_k^2 \underbrace{+ \frac{1}{2} I_{\frac{N'+1}{2}}^2}_{\text{only if } \frac{N'+1}{2} \in \mathbb{N}}\right.$$

$$\left. + 4 \sum_{\substack{l \neq m \\ 1 \leq l,m \leq N'}} s_l s_m I_l I_m - \sum_{k=1}^{N'} s_{2k} I_k I_{N'+1-k}\right). \quad (4.25)$$

(ii) *For any $\beta \neq 0$, $H_{0,\beta}^D(I)$ is nondegenerate at $I = 0$.*

CHAPTER 4. NONDEGENERACY AND CONVEXITY

Proof. To obtain the Birkhoff normal form (4.25) of H_V^D, we evaluate the formula (1.24) at $\alpha = 0$. To investigate the Hessian of $Q_{0,\beta}^D$ of $H_{0,\beta}^D(I)$ at $I = 0$, we again write

$$Q_{0,\beta}^D = \frac{2\beta}{16(N'+1)} \Delta^{N'} P^D \Delta^{N'}, \qquad (4.26)$$

similarly to (4.4), where $\Delta^{N'} = \mathrm{diag}\left(\sin\frac{k\pi}{2N'+2}\right)_{1 \leq k \leq N'}$, and where P^D is the $N' \times N'$-matrix which for N' even resp. odd is given by

$$\underbrace{\begin{pmatrix} 3 & 4 & \cdots & & & \cdots & 4 & 2 \\ 4 & 3 & 4 & \cdots & & \cdots & 4 & 2 & 4 \\ \vdots & & \ddots & & & \ddots & & \vdots \\ & & & 3 & 4 & 4 & 2 & & \\ & & & 4 & 3 & 2 & 4 & & \\ & & & 4 & 2 & 3 & 4 & & \\ & & & 2 & 4 & 4 & 3 & & \\ \vdots & & \ddots & & & \ddots & & \vdots \\ 4 & 2 & 4 & \cdots & & \cdots & 4 & 3 & 4 \\ 2 & 4 & \cdots & & & \cdots & & 4 & 3 \end{pmatrix}}_{(N' \text{ even})}, \quad \underbrace{\begin{pmatrix} 3 & 4 & \cdots & & \cdots & 4 & 2 \\ 4 & 3 & 4 & \cdots & \cdots & 4 & 2 & 4 \\ \vdots & & \ddots & & \ddots & & \vdots \\ & & & 3 & 4 & 2 & & \\ & & & 4 & 2 & 4 & & \\ & & & 2 & 4 & 3 & & \\ \vdots & & \ddots & & \ddots & & \vdots \\ 4 & 2 & 4 & \cdots & \cdots & 4 & 3 & 4 \\ 2 & 4 & \cdots & & \cdots & & 4 & 3 \end{pmatrix}}_{(N' \text{ odd})},$$

where we used that $s_{2k} = 2s_k c_k = 2s_k s_{N'+1-k}$ and, if $\frac{N'+1}{2} \in \mathbb{N}$, $s_{\frac{N'+1}{2}}^2 = \frac{1}{2}$. It follows that

$$\det Q_{0,\beta}^D = \left(\frac{2\beta}{16(N'+1)}\right)^{N-1} \cdot \det P^D \cdot \prod_{k=1}^{N'} \sin^2 \frac{k\pi}{2N'+2}. \qquad (4.27)$$

In order to see that P^D is regular, observe that $\det P^D \in \mathbb{Z}$. For N' even we show that $\det P^D \equiv 1 \mod 2$. Note that in this case the diagonal of P^D consists of 3's only, whereas all other products appearing in the Leibniz formula for $\det P^D$ contain at least one even number. Therefore $\det P^D \equiv 3^{N'} \mod 2 \equiv 1 \mod 2$. If N' is odd, the same argument shows that $\det P \equiv 2 \mod 4$. Hence, in both cases, $\det P^D \neq 0$ and hence if $\beta \neq 0$, $\det Q_{0,\beta}^D \neq 0$, and the nondegeneracy of the Hessian of $H_{0,\beta}^D(I)$ at $I = 0$ then follows by (4.27). □

We also need the full spectral information on the matrix P^D:

Lemma 4.2.2. *If N' is even, the eigenvalues of P^D are $4N' - 3$ (with multiplicity one), 1 (with multiplicity $\frac{N'}{2}$), and -3 (with multiplicity $\frac{N'}{2} - 1$). If N' is odd, the eigenvalues of P^D are 1 (with multiplicity $\frac{N'-1}{2}$), -3 (with multiplicity $\frac{N'-3}{2}$), and $\frac{1}{2}(4N' - 5)\left(1 \pm \sqrt{1 + \frac{8(4N'-1)}{(4N'-5)^2}}\right)$ (each with multiplicity one).*

Proof. We prove the lemma by providing the eigenvectors to the claimed eigenvalues. We have provided an analytical derivation of the eigenvalues of P^D in [45]. Let $(e_n)_{1 \leq n \leq N'}$ denote the standard basis in $\mathbb{R}^{N'}$.

4.2. DIRICHLET CHAINS

First assume that N' is even. Then $\sum_{j=1}^{N'} e_j$ (i.e. the vector $(1,\ldots,1)$) is an eigenvector to the eigenvalue $4N'-3$, $\frac{N'}{2}$ linearly independent eigenvectors to the eigenvalue 1 are given by $-e_1 + e_{N'}$, $-e_2 + e_{N'-1}$, ..., $-e_{N'/2} + e_{N'/2+1}$, and $\frac{N'}{2}-1$ linearly independent eigenvectors to the eigenvalue -3 are given by $e_1 - (e_{N'/2} + e_{N'/2+1}) + e_{N'}$, $e_2 - (e_{N'/2} + e_{N'/2+1}) + e_{N'-1}$, ..., $e_{N'/2-1} - (e_{N'/2} + e_{N'/2+1}) + e_{N'/2+2}$.

If N' is odd, intuitively, the eigenvectors of P^D of the even dimension $N'-1$ can be used to construct eigenvectors of P^D of dimension N' by "adding" an additional component in the "middle". Precisely, $\frac{N'-1}{2}$ linearly independent eigenvectors to the eigenvalue 1 are given by $-e_1 + e_{N'}$, $-e_2 + e_{N'-1}$, ..., $-e_{(N'-1)/2} + e_{(N'+1)/2}$, and $\frac{N'-3}{2}$ linearly independent eigenvectors to the eigenvalue -3 are given by $e_1 - (e_{(N'-1)/2} + e_{(N'+3)/2}) + e_{N'}$, $e_2 - (e_{(N'-1)/2} + e_{(N'+3)/2}) + e_{N'-1}$, ..., $e_{(N'-3)/2} - (e_{(N'-1)/2} + e_{(N'+3)/2}) + e_{(N'+5)/2}$. Finally, let

$$\lambda_\pm := \frac{1}{2}(4N'-5)\left(1 \pm \sqrt{1 + \frac{8(4N'-1)}{(4N'-5)^2}}\right);$$

then the two vectors $\sum_{j=1}^{N'} e_j + \left(-\frac{4N'-3}{4} + \frac{\lambda_\pm}{4}\right) e_{(N'+1)/2}$ are eigenvectors of P^D to the eigenvalue $\lambda_\pm$, respectively. All these facts can be checked by a direct computation. □

Lemma 4.2.3. *If $\beta < 0$, then $Q_{0,\beta}^D$ has $\lceil \frac{N'+1}{2} \rceil$ negative eigenvalues, whereas if $\beta > 0$, then $Q_{0,\beta}^D$ has $\lfloor \frac{N'-1}{2} \rfloor$ negative eigenvalues. In particular, for any $\beta \neq 0$, $Q_{0,\beta}^D$ is indefinite (and $H_{0,\beta}^D$ is therefore not convex). Moreover, for any $\beta \neq 0$, $H_{0,\beta}^D$ is directionally quasiconvex, but not quasiconvex at $I = 0$.*

Proof. Similarly to the proof of Lemma 4.1.3, we use the decomposition (4.26) of $Q_{0,\beta}^D$ to show that $Q_{0,\beta}^D$ can be continuously deformed to $\frac{2\beta}{16(N'+1)} P^D$: Consider for $0 \le t \le 1$

$$Q_{0,\beta}^D(t) := \frac{2\beta}{16(N'+1)} (t\Delta^{N'} + (1-t)\,\mathrm{Id})\, P^D\, (t\Delta^{N'} + (1-t)\,\mathrm{Id}).$$

As the diagonal matrix $t\Delta^{N'} + (1-t)\,\mathrm{Id}$ is positive definite for any $0 \le t \le 1$ and P^D is regular and symmetric, $Q_{0,\beta}^D(t)$ is a symmetric regular $N' \times N'$-matrix for any $0 \le t \le 1$. For $t=0$, $Q_{0,\beta}^D(0) = \frac{2\beta}{16(N'+1)} P^D$, whereas for $t=1$, $Q_{0,\beta}^D(1) = Q_{0,\beta}^D$. Therefore, the index (i.e. the number of negative eigenvalues) of $Q_{0,\beta}^D$ coincides with the index of $\frac{2\beta}{16(N'+1)} P^D$, or simply the index of P^D. The eigenvalues of P^D are listed in Lemma 4.2.2, and the claim of Lemma 4.2.3 now follows immediately. The quasiconvexity properties will be proved at the end of the proof of Proposition 4.2.4, after the proof of the corresponding statements for arbitrary $\alpha \neq 0$ (the proof of Proposition 4.2.4 does not use the quasiconvexity properties of $Q_{0,\beta}^D$). □

We now turn to the case $\alpha \neq 0$ and arbitrary $\beta \in \mathbb{R}$.

Proposition 4.2.4. *Assume that $\alpha \neq 0$ in (1.4). Then, for α fixed, $\det Q_{\alpha,\beta}^D$ is a polynomial in β of degree N' and has N' real zeroes (counted with multiplicities). When denoted by $\beta_k = \beta_k(\alpha)$ ($1 \leq k \leq N'$) and listed in increasing order, they satisfy*

$$\beta_1 \leq \ldots \leq \beta_{\lceil \frac{N'+1}{2} \rceil} < \alpha^2 < \beta_{\lceil \frac{N'+3}{2} \rceil} \leq \ldots \leq \beta_{N'}.$$

Moreover

$$\text{index}(Q_{\alpha,\beta}^D) = \begin{cases} \lceil \frac{N'+1}{2} \rceil & \text{for } \beta < \beta_1 \\ 0 & \text{for } \beta_{\lceil \frac{N'+1}{2} \rceil} < \beta < \beta_{\lceil \frac{N'+3}{2} \rceil} \\ \lfloor \frac{N'-1}{2} \rfloor & \text{for } \beta > \beta_{N'} \end{cases}$$

Hence $H_{\alpha,\beta}^D$ is convex if and only if $\beta_{\lceil \frac{N'+1}{2} \rceil} < \beta < \beta_{\lceil \frac{N'+3}{2} \rceil}$. Moreover, $H_{\alpha,\beta}^D$ is quasiconvex at $I = 0$ if and only if

$$\frac{\beta}{\alpha^2} \in \left(1 - \frac{2}{-1 + \sqrt{1 + 3\cos^2 \frac{\pi}{2N'+2}}}, 1 + \frac{2}{1 + \sqrt{1 + 3\cos^2 \frac{\pi}{2N'+2}}} \right), \quad (4.28)$$

and directionally quasiconvex at $I = 0$ for any $\beta \in \mathbb{R}$.

Proof. Fix $\alpha \in \mathbb{R} \setminus \{0\}$ and consider the map $\beta \mapsto \det(Q_{\alpha,\beta}^D)$. It follows from (1.24) that $\det(Q_{\alpha,\beta}^D)$ is a polynomial in β of degree at most N',

$$\det(Q_{\alpha,\beta}^D) = \sum_{j=0}^{N'} r_j \beta^j,$$

with $r_0 = \det(Q_{\alpha,0}^D)$ and $r_{N'} = \det(Q_{0,1}^D)$. By Proposition 4.2.1, $\det(Q_{0,1}^D) \neq 0$, hence the degree of the polynomial $\det(Q_{\alpha,\beta}^D)$ is precisely N'. We claim that $\det(Q_{\alpha,\beta}^D)$ has N' real zeroes (counted with multiplicities). For $|\beta|$ large enough, $\text{index}(Q_{\alpha,\beta}^D)$ is equal to $\text{index}(Q_{0,\beta}^D)$. By Lemma 4.2.3, $\text{index}(Q_{0,\beta}^D)$ is $\lfloor \frac{N'-1}{2} \rfloor$ for any $\beta > 0$ and $\lceil \frac{N'+1}{2} \rceil$ for any $\beta < 0$. Hence there exists some $R^D > 0$ such that $\text{index}(Q_{\alpha,\beta}^D) = \lfloor \frac{N'-1}{2} \rfloor$ for any $\beta > R^D$ and $\text{index}(Q_{\alpha,\beta}^D) = \lceil \frac{N'+1}{2} \rceil$ for any $\beta < -R^D$. For $\beta = \alpha^2$, Q_{α,α^2}^D is a positive multiple of the identity matrix, hence $\text{index}(Q_{\alpha,\alpha^2}^D) = 0$. It then follows that, when counted with multiplicities, $\text{index}(Q_{\alpha,\beta}^D)$ must change at least $\lceil \frac{N'+1}{2} \rceil$ times in the open interval $(-\infty, \alpha^2)$ and at least $\lfloor \frac{N'-1}{2} \rfloor$ times in (α^2, ∞). Since a change of $\text{index}(Q_{\alpha,\beta}^D)$ induces a real zero of $\det(Q_{\alpha,\beta}^D)$, our consideration shows that the function $\beta \mapsto \det(Q_{\alpha,\beta}^D)$ has N' real zeroes. Furthermore, we have $\beta_{\lceil \frac{N'+1}{2} \rceil}(\alpha) < \alpha^2 < \beta_{\lceil \frac{N'+3}{2} \rceil}(\alpha)$.

As in the odd periodic case, it is an open question, whether some variant of the KAM theorem requiring weaker nondegeneracy conditions than Kolmogorov's condition (2.9) can be applied to Dirichlet chains for these N' zeroes of $\beta \to \det(Q_{\alpha,\beta}^D)$.

4.2. DIRICHLET CHAINS

It remains to prove the quasiconvexity properties of $H^D_{\alpha,\beta}$ at $I = 0$. We again first prove the directional quasiconvexity. By the definition (2.17) of directional quasiconvexity, we have to show that $\xi = 0$ is the only solution in $\mathbb{R}^{N'}_{\geq 0}$ of the system of equations in (2.14). In the case at hand, $\omega_k = 2\sin\frac{k\pi}{N} = 2\sin\frac{k\pi}{2N'+2}$ for any $1 \leq k \leq N'$. Hence for any $\xi_1, \ldots, \xi_{N'} \geq 0$, $\langle \omega, \xi \rangle = 2\sum_{k=1}^{N'} s_k \xi_k > 0$ unless $\xi_1 = \ldots = \xi_{N'} = 0$. This proves the claim.

To prove the (non-directional) quasiconvexity properties, we have to investigate real solutions $\xi_1, \ldots, \xi_{N'}$ of (2.14) without the restriction $\xi \in \mathbb{R}^{N'}_{\geq 0}$. By (1.24), in this case the equations (2.14) are in detail given by

$$\sum_{k=1}^{N'} \left(-\gamma + 3s_k^2 + \frac{1}{2}\delta_{(2k,N'+1)}\right)\xi_k^2 + 4\sum_{1 \leq l \neq m \leq N'} s_l s_m \xi_l \xi_m - \sum_{l=1}^{N'} s_{2l}\xi_l \xi_{N'+1-l} = 0, \tag{4.29}$$

$$2\sum_{k=1}^{N'} s_k \xi_k = 0. \tag{4.30}$$

We again use a computation of the type (4.17) to substitute (4.30) into (4.29), obtaining the equation

$$\sum_{k=1}^{N'} \left(\gamma + s_k^2 - \frac{1}{2}\delta_{(2k,N'+1)}\right)\xi_k^2 + \sum_{k=1}^{N'} s_{2k}\xi_k \xi_{N'+1-k} = 0,$$

or, using that $c_k = s_{N'+1-k}$ for any $1 \leq k \leq N'$,

$$\sum_{k=1}^{\lfloor N'/2 \rfloor} ((\gamma+s_k^2)\xi_k^2 + (\gamma+c_k^2)\xi_{N'+1-k}^2 + 2s_{2k}\xi_k \xi_{N'+1-k}) + \underbrace{(\gamma+1)\xi_{(N'+1)/2}^2}_{\text{only if } \frac{N'+1}{2} \in \mathbb{N}} = 0. \tag{4.31}$$

We first consider the case N' even. Then the left hand side of (4.31) is the sum of $\frac{N'}{2}$ quadratic forms in the two variables $\xi_k, \xi_{N'+1-k}$ for $1 \leq k \leq \frac{N'}{2}$. The sum of these $\frac{N'}{2}$ quadratic forms has no nontrivial zero if and only if all of the quadratic forms are positive definite or all of them negative definite.

In other words, for any $1 \leq k \leq \frac{N'}{2}$, we have to investigate the definiteness of the quadratic form

$$q_k(x, y) := (\gamma + s_k^2)x^2 + (\gamma + c_k^2)y^2 + 2s_{2k}\, x\, y. \tag{4.32}$$

The two eigenvalues of the corresponding coefficient matrix

$$A_k := \begin{pmatrix} \gamma + s_k^2 & s_{2k} \\ s_{2k} & \gamma + c_k^2 \end{pmatrix} \tag{4.33}$$

have (strictly) the same sign if and only if

$$\gamma^2 + \gamma - \frac{3}{4}s_{2k}^2 > 0. \tag{4.34}$$

If the condition (4.34) is satisfied, the two eigenvalues are positive or negative if $2\gamma + 1$ is positive or negative, respectively. Hence, if (4.34) is fulfilled for any $1 \leq k \leq \frac{N'}{2}$, all N' eigenvalues of the $\frac{N'}{2}$ matrices $A_1, \ldots, A_{N'/2}$ defined by (4.33) have (strictly) the same sign, since $2\gamma + 1$ is independent of k. Observe that for any fixed γ, (4.34) is fulfilled for all $1 \leq k \leq \frac{N'}{2}$ if and only if it is true for $k = \frac{N'}{2}$, and that $s_{2(N'/2)}^2 = s_{N'}^2 = c_1^2 = \cos^2 \frac{\pi}{N}$. Alltogether, if N' is even, we arrive at the sufficient condition

$$\gamma^2 + \gamma - \frac{3}{4} \cos^2 \frac{\pi}{2N'+2} > 0 \qquad (4.35)$$

for the nonexistence of a nontrivial solution of the equation (4.31), i.e. for the quasiconvexity of $H_{\alpha,\beta}^D$ at $I = 0$.

If N' is odd, then, by the same kind of reasoning, we also arrive at the condition (4.35) for quasiconvexity. Note that the additional term $(\gamma+1)\xi_{(N'+1)/2}^2$ in the equation (4.31) does not influence the existence of nontrivial solutions: If we consider the equation (4.31) without the term $(\gamma+1)\xi_{(N'+1)/2}^2$ and find a nontrivial solution $(\xi_1, \ldots, \xi_{(N'-1)/2}, \xi_{(N'+3)/2}, \ldots, \xi_{N'})$, this solution can be extended to a nontrivial solution of the full equation (4.31) by setting $\xi_{(N'+1)/2} := 0$, and if such a solution to the reduced equation does not exist, then (4.35) holds, and the quadratic form $x \mapsto (\gamma+1)x^2$ has the same type of definiteness (positive or negative) as the sum of the $\frac{N'-1}{2}$ quadratic forms $(q_k)_{1 \leq k \leq (N'-1)/2}$ defined by (4.32).

Conversely, if (4.35) is not satisfied, i.e. if $\gamma^2 + \gamma - \frac{3}{4}\cos^2\frac{\pi}{2N'+2} \leq 0$, one easily constructs a nontrivial solution of (4.31) and then of (4.29)-(4.30).

We thus have proved that independently of the parity of N', $H_{\alpha,\beta}^D$ is quasiconvex at $I = 0$ if and only if (4.35) holds. This condition can be reformulated as

$$\gamma \notin \left[\frac{1}{2}\left(-1 - \sqrt{1 + 3\cos^2\frac{\pi}{2N'+2}}\right), \frac{1}{2}\left(-1 + \sqrt{1 + 3\cos^2\frac{\pi}{2N'+2}}\right) \right],$$

which then in terms of α, β leads to the claimed condition (4.28).

The case $\alpha = 0$ is equivalent to $\gamma = 0$, which can be explicitly checked by noting that by substituting (1.24) into the quasiconvexity condition (2.14), we obtain the system (4.29)-(4.30) for $\gamma = 0$. The above considerations then show that in this case, (4.29)-(4.30) does indeed have a nontrivial solution, which proves that for any $\beta \in \mathbb{R}$, $H_{0,\beta}^D$ is not quasiconvex at $I = 0$. This completes the proof of Proposition 4.2.4, and also of Lemma 4.2.3. $\square$

Proof of Theorem 1.3.3. Part (i) follows from by Proposition 4.2.4, and (ii) follows from Propostion 4.2.1 and Lemma 4.2.3. $\square$

Chapter 5

The Foliation of the Phase Space by the Moment Map of an Integrable Approximation of Even Periodic Chains

In this chapter we describe the geometry of the moment map of the truncated resonant normal form (1.16) of even periodic FPU chains potential V whose expansion (1.4) satisfies $(\alpha, \beta) \neq (0,0)$. The case $\beta = \alpha^2$ is special as in this case the resonant term $R_{\alpha,\beta}$ vanishes, and the normal form (1.16) is the Birkhoff normal form of order four of the periodic Toda lattice. Its foliation is well known - it is the one of uncoupled harmonic oscillators. Hence we will investigate the case $\beta \neq \alpha^2$ only. The special case $\alpha = 0$, i.e. the β-chain, has been partially studied by Rink [83], and it turns out that many of his results continue to hold in the general case. This seems to be surprising, but if we recall that the β-chain is more historically than structurally special, it seems plausible that we find the qualitative behaviour of this special case also in the general case.

Using the notation $\tilde{k} \equiv \tilde{k}(k) = \frac{N}{2} - k$, the integrals of Theorem 1.2.5 can be grouped as

$$(\mathcal{H}_k, \mathcal{F}_{\tilde{k}}, G_k, K_k)_{1 \leq k < \frac{N}{4}}, \quad I_{\frac{N}{2}}, \quad \mathcal{H}_{\frac{N}{4}}, K_{\frac{N}{4}}, \tag{5.1}$$

where for $1 \leq k < \frac{N}{2}$

$$\mathcal{H}_k := I_k + I_{N-k}, \quad G_k := I_k - I_{\tilde{k}}.$$

(Here we used that $G_k = (I_k + I_{\frac{N}{2}+k}) - (I_{\frac{N}{2}-k} + I_{\frac{N}{2}+k})$ is the difference of two of the integrals listed in Theorem 1.2.5.)

Using the assumption $\beta \neq \alpha^2$, we rewrite the integrals $(K_l)_{1 \leq l \leq \frac{N}{4}}$ as follows. Throughout this chapter, we continue to use the bifurcation parameter

$$\gamma \equiv \gamma(\alpha, \beta) := \frac{\alpha^2}{\alpha^2 - \beta}. \tag{5.2}$$

introduced in (4.10). Further note that the quantity d_k^- appearing in the definition (1.19) of the integrals $(K_k)_{1 \leq k < \frac{N}{4}}$ can be rewritten as

$$d_k^- = -\alpha^2 + (\beta - \alpha^2) s_k^2 = (\beta - \alpha^2)(\gamma + s_k^2). \tag{5.3}$$

Using (5.3), one has for any $1 \leq k < \frac{N}{4}$ by (1.19)-(1.21) (note that $s_{\tilde{k}} = c_k$ and $s_{2\tilde{k}} = s_{2k}$)

$$K_k = -d_k^-(J_k^2 + M_k^2) - d_{\tilde{k}}^-(J_{\tilde{k}}^2 + M_{\tilde{k}}^2) + 2(\beta - \alpha^2) s_{2k}(J_k J_{\tilde{k}} - M_k M_{\tilde{k}})$$
$$= s_{2k}(\alpha^2 - \beta) \left(\frac{\gamma + s_k^2}{s_{2k}} (M_k^2 + J_k^2) + \frac{\gamma + c_k^2}{s_{2k}} (M_{\tilde{k}}^2 + J_{\tilde{k}}^2) + 2(M_k M_{\tilde{k}} - J_k J_{\tilde{k}}) \right), \tag{5.4}$$

and for $k = \frac{N}{4}$,

$$K_k = \alpha^2 J_k^2 - (\beta - 2\alpha^2) M_k^2 = (\alpha^2 - \beta) \left(\gamma J_k^2 + (1 + \gamma) M_k^2 \right).$$

In the sequel, we will for simplicity *omit* the factors $s_{2k}(\alpha^2 - \beta)$ in the integrals $(K_k)_{1 \leq k \leq \frac{N}{4}}$, since they do not influence the geometry of the level sets of the K_k's.

As already mentioned in the proof of Lemma 3.1.9, each of the $\lfloor \frac{N}{4} \rfloor + 1$ groups of integrals listed in (5.1) depends only on a subset of the variables $\{(x_k, y_k)_{1 \leq k \leq N-1}\}$, and these subsets form a disjoint partition of the set of all variables $\{(x_k, y_k)_{1 \leq k \leq N-1}\}$. More precisely, we have the following result.

Proposition 5.0.5. *The phase space $T^* \mathbb{R}^{N-1}$ of the truncated resonant normal form H_V^{trunc} given by (1.16) is the direct sum of invariant symplectic subspaces,*

$$T^* \mathbb{R}^{N-1} = \bigoplus_{0 \leq k \leq \frac{N}{4}} \mathcal{P}_k,$$

where

$$\mathcal{P}_k = \{(x_j, y_j)_{1 \leq j \leq N-1} \in T^* \mathbb{R}^{N-1} | x_j = y_j = 0 \ \forall j \notin \{k, N-k, \tilde{k}, N-\tilde{k}\}\}.$$

The foliation of $T^* \mathbb{R}^{N-1}$ by level sets of the integrals (5.1) is the Cartesian product of the foliations of the $\mathcal{P}_k$ by level sets of those integrals which depend on x_k, y_k, x_{N-k}, y_{N-k}, $x_{\tilde{k}}$, $y_{\tilde{k}}$, $x_{N-\tilde{k}}$, and $y_{N-\tilde{k}}$.

We now analyze the foliations of these $\lfloor \frac{N}{4} \rfloor + 1$ subspaces $\mathcal{P}_0$, $\mathcal{P}_{\frac{N}{4}}$, and $(\mathcal{P}_k)_{0 < k < \frac{N}{4}}$ separately. Note that $\mathcal{P}_0$ and $\mathcal{P}_k$ ($0 < k < \frac{N}{4}$) can be canonically identified with $T^* \mathbb{R}$ and $T^* \mathbb{R}^4$, respectively, and $\mathcal{P}_{\frac{N}{4}}$ with $T^* \mathbb{R}^2$ (if $\frac{N}{4} \in \mathbb{N}$) or $\{0\}$ (if $\frac{N}{4} \notin \mathbb{N}$).

5.1 Foliation of $\mathcal{P}_0$

The integral $I_{\frac{N}{2}}$ foliates $T^*\mathbb{R}$ by circles, centered at the origin.

5.2 Foliation of $\mathcal{P}_{\frac{N}{4}}$ for $\frac{N}{4} \in \mathbb{Z}$

In this section we study the geometry of the moment map $\mathcal{M} : T^*\mathbb{R}^2 \to \mathbb{R}^2$ defined by the integrable system with commuting integrals $H \equiv \mathcal{H}_{\frac{N}{4}}$ and $K \equiv K_{\frac{N}{4}}$. For convenience we introduce the following notation. We denote the standard coordinates of $T^*\mathbb{R}^2$ by $(x,y) = (x_1, x_2, y_1, y_2)$ and introduce the action variables $I_j = \frac{1}{2}(x_j^2 + y_j^2)$ ($j = 1, 2$), as well as the Hopf variables M, J, L as given by (1.10),

$$(M, J, L) = \frac{1}{2}(x_1 y_2 - x_2 y_1, x_1 x_2 + y_1 y_2, I_1 - I_2).$$

As already remarked in (1.12), we have

$$M^2 + J^2 + L^2 = H^2. \tag{5.5}$$

The moment map $\mathcal{M} = (H, K)$ then takes the form

$$H = \frac{1}{2}(I_1 + I_2), \quad K = (1 + \gamma)M^2 + \gamma J^2.$$

Furthermore, we may replace K by K_γ, given by

$$K_\gamma := \begin{cases} (1+\gamma)M^2 + \gamma J^2 & \gamma \notin \{-1, 0\}, \\ M & \gamma = 0, \\ J & \gamma = -1. \end{cases} \tag{5.6}$$

We first observe that the origin $(x,y) = (0,0)$ of $T^*\mathbb{R}^2$ is the only critical point of the moment map $\mathcal{M}$ with rank $d_{(x,y)}\mathcal{M} = 0$. Moreover, it follows immediately from $H = \frac{1}{2}(I_1 + I_2)$ that for any $\gamma \in \mathbb{R}$,

$$\mathcal{M}^{-1}\{(0,0)\} = \{(0,0)\}.$$

To analyze the critical points $(x,y) \in T^*\mathbb{R}^2 \setminus \{(0,0)\}$ with rank $d_{(x,y)}\mathcal{M} = 1$, we perform a symplectic reduction via the Hamiltonian vector field of H. On the sphere $\mathbb{S}^3_\rho = \{H = \rho^2/4\}$ of radius $\rho > 0$ in $T^*\mathbb{R}^2$ we define the Hopf map

$$\mathcal{F} : \mathbb{S}^3_\rho \to \mathbb{S}^2_r, \quad (x, y) \mapsto (M, J, L), \tag{5.7}$$

where $r = \sqrt{M^2 + J^2 + L^2}|_{\mathbb{S}^3_\rho} = H|_{\mathbb{S}^3_\rho} = \frac{\rho^2}{4}$. The fibers of $\mathcal{F}$ are circles obtained by the $\mathbb{S}^1$-action of H. The reduced system is then given by $(\mathbb{S}^2_r, X_\gamma)$, where X_γ denotes the reduced Hamiltonian vector field induced by K_γ. To compute X_γ, we claim that the equations of motion in the reduced system corresponding to the Hamiltonian K_γ are given by

$$\frac{d}{dt} \begin{pmatrix} M \\ J \\ L \end{pmatrix} = \begin{pmatrix} M \\ J \\ L \end{pmatrix} \times \begin{pmatrix} \partial_M K_\gamma \\ \partial_J K_\gamma \\ \partial_L K_\gamma \end{pmatrix}. \tag{5.8}$$

Indeed, following the procedure of reduction in section I.5 of [16], the formula (5.8) follows from

$$\dot{w}_j = \{w_j, K_\gamma\} = \sum_{i=1}^{3} \frac{\partial K_\gamma}{\partial w_i} \{w_j, w_i\} \qquad (1 \leq j \leq 3) \qquad (5.9)$$

and the commutation relations among the variables $(w_1, w_2, w_3) = (M, J, L)$ stated in Lemma 3.1.8. We then get by (5.6) and (5.8)

$$X_\gamma = \begin{cases} (-2\gamma JL, 2(1+\gamma)ML, -2MJ) & \gamma \notin \{-1, 0\}, \\ (0, L, -J) & \gamma = 0, \\ (-L, 0, M) & \gamma = -1. \end{cases}$$

The foliation of $\mathbb{S}_r^2$ by level sets of K_γ depends on the bifurcation parameter γ. In the case $\gamma = 0$, $(\pm r, 0, 0)$ are the only two fixed points of X_0. They are both elliptic, and the level sets of K_0 in $\mathbb{S}_r^2 \setminus \{(\pm r, 0, 0)\}$ are circles. Similarly, in the case $\gamma = -1$, $(0, \pm r, 0)$ are the only two fixed points of X_{-1}. They are both elliptic, and the level sets of K_{-1} in $\mathbb{S}_r^2 \setminus \{(0, \pm r, 0)\}$ are circles. It remains to consider the case $\gamma \notin \{-1, 0\}$. Then X_γ admits six fixed points,

$$(\pm r, 0, 0), \quad (0, \pm r, 0), \quad (0, 0, \pm r), \qquad (5.10)$$

two of which are hyperbolic and the remaining four elliptic. Note that the corresponding values of K_γ for the three pairs (5.10) of points are $(1+\gamma)r^2$, γr^2, and 0, respectively, and that the two hyperbolic fixed points are contained in the same connected component of the inverse image of K_γ in $\mathbb{S}_r^2$. This component consists of two great circles consisting of four half circles, each of which is a heteroclinic X_γ-orbit connecting the two hyperbolic fixed points.

	hyperbolic fixed points	critical value
$\gamma < -1$	$(\pm r, 0, 0)$	$(1+\gamma)r^2$
$-1 < \gamma < 0$	$(0, 0, \pm r)$	0
$\gamma > 0$	$(0, \pm r, 0)$	γr^2

(5.11)

Let us verify the claimed classification of the two fixed points $(0, 0, \varepsilon r)$ with $\varepsilon \in \{\pm\}$; the other four fixed points can be treated similarly. Near $(0, 0, \varepsilon r)$ we choose M, J as coordinates of $\mathbb{S}_r^2$. The equations of motion induced by X_γ on $\mathbb{S}_r^2$ in these coordinates are given by

$$\dot{M} = -\varepsilon 2\gamma J \sqrt{r^2 - M^2 - J^2},$$
$$\dot{J} = \varepsilon 2(1+\gamma) M \sqrt{r^2 - M^2 - J^2}.$$

If linearized at $(0, 0, \varepsilon r)$, the corresponding linear system is given by the 2×2-matrix εA with

$$A = 2r \begin{pmatrix} 0 & -\gamma \\ 1+\gamma & 0 \end{pmatrix}.$$

5.3. FOLIATION OF $\mathcal{P}_K$ FOR $0 < K < \frac{N}{4}$ 69

The eigenvalues of A are given by the equation $\lambda^2 + 4\gamma(1+\gamma)r^2 = 0$, i.e.

$$\lambda_{1,2} = \pm 2r\sqrt{-\gamma(1+\gamma)}.$$

As $-\gamma(1+\gamma) > 0$ if and only if $-1 < \gamma < 0$, it follows that for $-1 < \gamma < 0$, $\lambda_{1,2}$ are real and nonzero and hence that $(0, 0, \pm r)$ are both hyperbolic fixed points, whereas they are both elliptic if $\gamma < -1$ or $\gamma > 0$. For $-1 < \gamma < 0$, the inverse image $K_\gamma^{-1}(\{0\})$ is given by

$$K_\gamma^{-1}(\{0\}) = \{(M, J, L) | (1+\gamma)M^2 + \gamma J^2 = 0; M^2 + J^2 + L^2 = r^2\}$$
$$= \left\{(M, J, L) \Big| M = \pm\sqrt{\left|\frac{\gamma}{1+\gamma}\right|} J; M^2 + J^2 + L^2 = r^2\right\},$$

whereas for $\gamma < -1$ or $\gamma > 0$, we have $K_\gamma^{-1}(\{0\}) = \{(0, 0, \pm r)\}$.

5.3 Foliation of $\mathcal{P}_k$ for $0 < k < \frac{N}{4}$

In this section we present a detailed study of the geometry of the moment map $\mathcal{M} : T^*\mathbb{R}^4 \to \mathbb{R}^4$ given by the integrable system $\mathcal{P}_k$ with commuting integrals $\mathcal{H}_k, \mathcal{H}_{\bar{k}}, G_k, K_k$ for any $1 \le k < \frac{N}{4}$. As before we restrict the discussion to FPU chains with parameters α, β satisfying $\beta \ne \alpha^2$. We show that in this case the vector field induced by K_k exhibits hyperbolic dynamics.

For convenience we introduce the following notation, similarly to the notation of section 5.2. We denote the standard coordinates of $T^*\mathbb{R}^4$ by (x, y) with $x = (x_i)_{1 \le i \le 4}$ and $y = (y_i)_{1 \le i \le 4}$ and introduce the action variables $I_j = \frac{1}{2}(x_j^2 + y_j^2)$ ($1 \le j \le 4$), as well as the Hopf variables $(M_i, J_i, L_i)_{1 \le i \le 2}$, given by

$$(M_1, J_1, L_1) = \frac{1}{2}(x_1 y_2 - x_2 y_1, x_1 x_2 + y_1 y_2, I_1 - I_2),$$
$$(M_2, J_2, L_2) = \frac{1}{2}(x_3 y_4 - x_4 y_3, x_3 x_4 + y_3 y_4, I_3 - I_4).$$

By Lemma 3.1.8, the Poisson brackets between the variables $(M_i, J_i, L_i)_{1 \le i \le 2}$ are given by

$$\{M_i, J_i\} = -L_i, \quad \{J_i, L_i\} = -M_i, \quad \{L_i, M_i\} = -J_i,$$

whereas all other brackets vanish, and as in (5.5), we have $M_i^2 + J_i^2 + L_i^2 = H_i^2$ for $1 \le i \le 2$.

The moment map $\mathcal{M}$ then takes the form

$$\mathcal{M} : T^*\mathbb{R}^4 \to \mathbb{R}^4, \quad (x, y) \mapsto (H_1, H_2, G, K_\gamma)$$

with

$$H_1 = \frac{1}{2}(I_1 + I_2); \quad H_2 = \frac{1}{2}(I_3 + I_4); \quad G = L_1 - L_2 \qquad (5.12)$$

and where K_γ is a scalar multiple of K_k (see (5.4)), given by

$$K_\gamma = \sum_{i=1}^{2} \frac{1}{2} d_{i,\gamma}(M_i^2 + J_i^2) + (M_1 M_2 - J_1 J_2) \tag{5.13}$$

with the coefficients $d_{1,\gamma}$, $d_{2,\gamma}$ defined by

$$d_{1,\gamma} = \frac{\gamma + s_k^2}{s_{2k}}, \quad d_{2,\gamma} = \frac{\gamma + c_k^2}{s_{2k}}.$$

(The definition of the integral G above differs from the one given in (5.1) by the integral $H_1 - H_2$ as $I_1 - I_3 = L_1 - L_2 + H_1 - H_2$.)

First note that the origin $(0,0)$ in $T^*\mathbb{R}^4$ is the only critical point of $\mathcal{M}$ with rank zero. Moreover, by $H_1 = \frac{1}{2}(I_1 + I_2)$ and $H_2 = \frac{1}{2}(I_3 + I_4)$, $\mathcal{M}^{-1}\{(0,0)\} = \{(0,0)\}$. The next observation is that when restricted to $T^*\mathbb{R}^2 \times \{0\}$, one has $G = \frac{1}{2}(I_1 - I_2)$ and $K_\gamma = d_{1,\gamma}(H_1^2 - L_1^2)$, hence G and K_γ are functions of I_1, I_2 alone, and $\mathcal{M}|_{T^*\mathbb{R}^2 \times \{0\}}$ may therefore be replaced by the map $(x,y) \mapsto (I_1, I_2, 0, 0)$. The geometry of the latter map is the one of two uncoupled harmonic oscillators. The subspace $\{0\} \times T^*\mathbb{R}^2$ can be treated similarly. It remains to study the restriction of $\mathcal{M}$ to $T^*\mathbb{R}^4 \setminus ((T^*\mathbb{R}^2 \times \{0\}) \cup (\{0\} \times T^*\mathbb{R}^2))$. The Hamiltonian vector fields of H_1 and H_2 induce a torus action on $T^*\mathbb{R}^2$. Analogously to the Hopf map (5.7), the corresponding symplectic reduction is given by the product of two Hopf maps,

$$\mathcal{F}: \mathbb{S}_{\rho_1}^3 \times \mathbb{S}_{\rho_2}^3 \to \mathbb{S}_{r_1}^2 \times \mathbb{S}_{r_2}^2, \; (x,y) \mapsto (M_i, J_i, L_i)_{1 \leq i \leq 2}$$

where for $i = 1, 2$, $\mathbb{S}_{\rho_i}^3 = \{H_i = \rho_i^2/4\}$ is a sphere in $T^*\mathbb{R}^2$ and $r_i = \rho_i^2/4 = \sqrt{M_i^2 + J_i^2 + L_i^2}|_{\mathbb{S}_{\rho_i}^3} = H_i|_{\mathbb{S}_{\rho_i}^3}$. The fibers of $\mathcal{F}$ are 2-dimensional tori, obtained by the $\mathbb{S}^1 \times \mathbb{S}^1$-action of $H_1 \cdot H_2$. The reduced system is then given by $(\mathbb{S}_{r_1}^2 \times \mathbb{S}_{r_2}^2, Y, X_\gamma)$, where Y and X_γ denote the reduced Hamiltonian vector fields induced by G and K_γ, respectively. To compute Y and X_γ, note that the equations of motion in the reduced system, corresponding to a Hamiltonian H, are given by

$$\frac{d}{dt}\begin{pmatrix} M_i \\ J_i \\ L_i \end{pmatrix} = \begin{pmatrix} M_i \\ J_i \\ L_i \end{pmatrix} \times \begin{pmatrix} \partial_{M_i} H \\ \partial_{J_i} H \\ \partial_{L_i} H \end{pmatrix}, \quad i = 1, 2 \tag{5.14}$$

- see section 5.2, in particular formula (5.9), for details of this procedure. We then get

$$Y = \begin{pmatrix} J_1 \\ -M_1 \\ 0 \\ -J_2 \\ M_2 \\ 0 \end{pmatrix}, \quad X_\gamma = \begin{pmatrix} (J_2 - d_{1,\gamma} J_1) L_1 \\ (d_{1,\gamma} M_1 + M_2) L_1 \\ -(M_1 J_2 + M_2 J_1) \\ (J_1 - d_{2,\gamma} J_2) L_2 \\ (d_{2,\gamma} M_2 + M_1) L_2 \\ -(M_1 J_2 + M_2 J_1) \end{pmatrix}. \tag{5.15}$$

5.3. FOLIATION OF $\mathcal{P}_K$ FOR $0 < K < \frac{N}{4}$

Further we introduce the reduced moment map

$$\mathcal{M}_\gamma : \mathbb{S}^2_{r_1} \times \mathbb{S}^2_{r_2} \to \mathbb{R}^2, \ (M_i, J_i, L_i)_{1 \le i \le 2} \mapsto (G, K_\gamma).$$

We now study the critical points of $\mathcal{M}_\gamma$ with rank zero, i.e. points of $\mathbb{S}^2_{r_1} \times \mathbb{S}^2_{r_2}$ which are fixed points of both Y and X_γ. From the expressions for Y and X_γ given by (5.15), one easily sees that there are only four such critical points, namely

$$(M_i, J_i, L_i)_{1 \le i \le 2} = \varepsilon(0, 0, r_1, 0, 0, \pm r_2),$$

where $\varepsilon \in \{\pm\}$. Choose $(M_i, J_i)_{1 \le i \le 2}$ as coordinates of $\mathbb{S}^2_{r_1} \times \mathbb{S}^2_{r_2}$ near these points $\varepsilon(0, 0, r_1, 0, 0, \pm r_2)$. The equations of motion are then given by

$$(\dot{M}_1, \dot{J}_1) = ((J_2 - d_{1,\gamma} J_1) L_1, (d_{1,\gamma} M_1 + M_2) L_1)$$

and

$$(\dot{M}_2, \dot{J}_2) = ((J_1 - d_{2,\gamma} J_2) L_2, (d_{2,\gamma} M_2 + M_1) L_2),$$

where $L_i = \pm r_i \sqrt{1 - (M_i^2 + J_i^2)/r_i^2}$. If linearized at $(0, 0, \xi, 0, 0, \eta)$ for $\xi \in \{\pm r_1\}$, $\eta \in \{\pm r_2\}$, the corresponding linear system is given by the 4×4-matrix $A \equiv A_{\xi, \eta}$,

$$A = \begin{pmatrix} 0 & -d_{1,\gamma}\xi & 0 & \xi \\ d_{1,\gamma}\xi & 0 & \xi & 0 \\ 0 & \eta & 0 & -d_{2,\gamma}\eta \\ \eta & 0 & d_{2,\gamma}\eta & 0 \end{pmatrix}. \tag{5.16}$$

We first turn to the critical points $\varepsilon(0, 0, r_1, 0, 0, -r_2)$; their values by $\mathcal{M}_\gamma$ are $(\varepsilon(r_1 + r_2), 0)$, and

$$\mathcal{M}_\gamma^{-1}\{(\varepsilon(r_1 + r_2), 0)\} = \{\varepsilon(0, 0, r_1, 0, 0, -r_2)\}.$$

We will prove the following proposition together with the analogous statement on the other two fixed points of X_γ, i.e. Proposition 5.3.2.

Proposition 5.3.1. *For any $\gamma \in \mathbb{R}$, $1 \le k < \frac{N}{4}$, and $\varepsilon \in \{\pm\}$, the critical point $\varepsilon(0, 0, r_1, 0, 0, -r_2)$ is a (possibly degenerate) elliptic fixed point of X_γ.*

The values of the other two critical points $\varepsilon(0, 0, r_1, 0, 0, r_2)$ by $\mathcal{M}_\gamma$ are $(\varepsilon(r_1 - r_2), 0)$. The inverse image of $(\varepsilon(r_1 - r_2), 0)$ might have several connected components, depending on the values of γ and the additional bifurcation parameter

$$r := \frac{r_1}{r_2} > 0.$$

(The third bifurcation parameter is the index k, $1 \le k \le \frac{N}{4}$.) We restrict the discussion to the case $r > 1$; the case $r > 1$ can be treated analogously and does not lead to any qualitatively new insights. Our results on the critical points $\varepsilon(0, 0, r_1, 0, 0, r_2)$ are collected in Theorem 1.4.1, the main result of this chapter. To prove Theorem 1.4.1 we first prove the hyperbolicity criterion (1.25),

$$\left| (\gamma + s_k^2)\sqrt{r} + (\gamma + c_k^2)\frac{1}{\sqrt{r}} \right| < 2s_{2k}. \tag{5.17}$$

Let us first reformulate (1.25) in terms of the function

$$f : (0, \sqrt{r}] \to \mathbb{R}, \ q \mapsto (\gamma + s_k^2)q + (\gamma + c_k^2)\frac{1}{q}. \tag{5.18}$$

Proposition 5.3.2. *Let $\gamma \in \mathbb{R}$ be arbitrary and assume that $1 \leq k < \frac{N}{4}$ and $0 < r \leq 1$. Then the following statements hold:*

(i) If $|f(\sqrt{r})| \geq 2s_{2k}$, then $\pm(0, 0, r_1, 0, 0, r_2)$ are (possibly degenerate) elliptic fixed points of X_γ.

(ii) If $|f(\sqrt{r})| < 2s_{2k}$, then $\pm(0, 0, r_1, 0, 0, r_2)$ are hyperbolic fixed points of X_γ. Their stable and and unstable manifolds have each dimension two.

Proof. We compute $\det(\lambda - A) = \det(A - \lambda)$ for the matrix A given by (5.16) and get

$$\det(A - \lambda) = \lambda^4 + a\lambda^2 + b^2, \tag{5.19}$$

where

$$a = d_{1,\gamma}^2 \xi^2 + d_{2,\gamma}^2 \eta^2 - 2\xi\eta \tag{5.20}$$

and

$$b = (d_{1,\gamma} d_{2,\gamma} - 1)\xi\eta. \tag{5.21}$$

The discriminant of (5.19) is given by

$$a^2 - 4b^2 = (a - 2b)(a + 2b)$$
$$= (d_{1,\gamma}^2 \xi^2 - 2d_{1,\gamma} d_{2,\gamma} \xi\eta + d_{2,\gamma}^2 \eta^2)(d_{1,\gamma}^2 \xi^2 + 2d_{1,\gamma} d_{2,\gamma} \xi\eta + d_{2,\gamma}^2 \eta^2 - 4\xi\eta).$$

We first consider the case $\xi\eta = -r_1 r_2$, i.e. we turn to the proof of Proposition 5.3.1. Then

$$a^2 - 4b^2 = \underbrace{(d_{1,\gamma} r_1 + d_{2,\gamma} r_2)^2}_{\geq 0} \underbrace{\left((d_{1,\gamma} r_1 - d_{2,\gamma} r_2)^2 + 4r_1 r_2\right)}_{> 0} \geq 0$$

and

$$a = d_{1,\gamma}^2 r_1^2 + d_{2,\gamma}^2 r_2^2 + 2r_1 r_2 > 0.$$

Hence $\lambda^2 = \mu_\pm$ with

$$\mu_\pm = -\frac{a}{2}\left(1 \pm \sqrt{1 - \left(\frac{2b}{a}\right)^2}\right) \leq 0. \tag{5.22}$$

More precisely, one always has $\mu_+ < 0$ whereas $\mu_- = 0$ if and only if $b = 0$. Hence

$$\lambda_{1,2,3,4} = \pm i \sqrt{\frac{a}{2}}\left(1 \pm \sqrt{1 - \left(\frac{2b}{a}\right)^2}\right)^{\frac{1}{2}} \in i\mathbb{R}.$$

5.3. FOLIATION OF $\mathcal{P}_K$ FOR $0 < K < \frac{N}{4}$

It follows that the fixed points $\pm(0, 0, r_1, 0, 0, -r_2)$ of X_γ are both elliptic, except in the case $b = 0$ (i.e. $d_{1,\gamma} d_{2,\gamma} = 1$, or $\gamma^2 + \gamma - \frac{3}{4} s_{2k}^2 = 0$) where they are degenerate elliptic. This proves Proposition 5.3.1.

Let us now turn to the case $\xi\eta = r_1 r_2$, i.e. Proposition 5.3.2. Then

$$a^2 - 4b^2 = (d_{1,\gamma} r_1 - d_{2,\gamma} r_2)^2 ((d_{1,\gamma} r_1 + d_{2,\gamma} r_2)^2 - 4 r_1 r_2) \quad (5.23)$$

$$= \left(\frac{r_1 r_2}{s_{2k}}\right)^2 \left(d_{1,\gamma} \sqrt{r} - d_{2,\gamma} \frac{1}{\sqrt{r}}\right)^2 \left(f(\sqrt{r})^2 - 4 s_{2k}^2\right), \quad (5.24)$$

where we recall from (5.18) the definition of $f(q)$. First we have to establish some auxiliary results

Lemma 5.3.3. *Let $\gamma \in \mathbb{R}$ be arbitrary and assume that $1 \leq k < \frac{N}{4}$ and $0 < r < 1$. Then $d_{1,\gamma} \sqrt{r} - d_{2,\gamma} \frac{1}{\sqrt{r}} \neq 0$.*

Proof. Assume that $d_{1,\gamma} \sqrt{r} - d_{2,\gamma} \frac{1}{\sqrt{r}} = 0$. First note that if $d_{1,\gamma} = 0$, then $d_{2,\gamma} = 0$. Thus $d_{1,\gamma} = d_{2,\gamma}$ or $\gamma + s_k^2 = \gamma + c_k^2$. As $1 \leq k < \frac{N}{4}$ by assumption, this leads to a contradiction. Hence $d_{1,\gamma} \neq 0$, and $d_{1,\gamma} \sqrt{r} - d_{2,\gamma} \frac{1}{\sqrt{r}} = 0$ is equivalent to $r = \frac{d_{2,\gamma}}{d_{1,\gamma}}$. As $r \leq 1$ by assumption, we conclude that $d_{2,\gamma} \leq d_{1,\gamma}$, or $c_k^2 \leq s_k^2$, which in turn contradicts the assumption $1 \leq k < \frac{N}{4}$. $\square$

Note that in the case $\xi\eta = r_1 r_2$, the parameter a, given by (5.20), can be rewritten as

$$a = d_{1,\gamma}^2 r_1^2 + d_{2,\gamma}^2 r_2^2 - 2 r_1 r_2$$

$$= \frac{r_1 r_2}{s_{2k}^2}\left((\gamma + s_k^2)^2 r + (\gamma + c_k^2)^2 \frac{1}{r} - 2 s_{2k}^2\right) \quad (5.25)$$

Lemma 5.3.4. *Let $\gamma \in \mathbb{R}$ be arbitrary and assume that $1 \leq k < \frac{N}{4}$ and $r > 0$. If $a < 0$, then*

$$|f(\sqrt{r})| < 2 s_{2k}.$$

Proof. We argue indirectly and assume that for some γ_0, k, and r, we have $a < 0$ but $|f(\sqrt{r})| \geq 2 s_{2k}$. The latter inequality can be written as

$$g_{\gamma_0}(r) := (\gamma_0 + s_k^2)^2 r + (\gamma_0 + c_k^2)^2 \frac{1}{r} \geq \frac{7}{2} s_{2k}^2 - 2 \gamma_0^2 - 2 \gamma_0. \quad (5.26)$$

In view of (5.25), the inequality $a < 0$ can be expressed as $g_{\gamma_0}(r) < 2 s_{2k}^2$. Together with (5.26), this leads to the two inequalities

$$\frac{7}{2} s_{2k}^2 - 2 \gamma_0^2 - 2 \gamma_0 \leq g_{\gamma_0}(r) < 2 s_{2k}^2. \quad (5.27)$$

If $\gamma_0 = -s_k^2$ or $\gamma_0 = -c_k^2$, we have $\frac{7}{2} s_{2k}^2 - 2\gamma_0^2 - 2\gamma_0 = 4 s_{2k}^2$, directly contradicting (5.27). If $\gamma_0 \notin \{-s_k^2, -c_k^2\}$, then $g_{\gamma_0}(q) \to \infty$ for $q \to 0$ or $q \to \infty$. Further, for any $\gamma \in \mathbb{R}\setminus\{-s_k^2, -c_k^2\}$, g_γ achieves its minimum on $\mathbb{R}_{>0}$ at $q = (\gamma+c_k^2)/(\gamma+s_k^2)$, and

$$\min_{q>0} g_\gamma(q) = 2\gamma^2 + 2\gamma + \frac{1}{2}s_{2k}^2. \tag{5.28}$$

Note that $\gamma \mapsto \min_{q>0} g_\gamma(q) = 2\gamma^2 + 2\gamma + \frac{1}{2}s_{2k}^2$ is a continuous function on $\mathbb{R}$, tending to ∞ as $\gamma \to \infty$. Hence there exists $\gamma_1 \in \mathbb{R}$ such that

$$g_{\gamma_0}(r) = \min_{q>0} g_{\gamma_1}(q). \tag{5.29}$$

In particular, $\min_{q>0} g_{\gamma_1}(q) \geq \min_{q>0} g_{\gamma_0}(q)$, i.e. $2\gamma_1^2 + 2\gamma_1 \geq 2\gamma_0^2 + 2\gamma_0$. When combined with (5.27) and (5.29), the latter inequality leads to

$$\frac{7}{2}s_{2k}^2 - 2\gamma_1^2 - 2\gamma_1 \leq \frac{7}{2}s_{2k}^2 - 2\gamma_0^2 - 2\gamma_0$$
$$\leq g_{\gamma_0}(r) = \min_{q>0} g_{\gamma_1}(q).$$

If $\min_{q>0} g_\gamma(q) \geq \frac{7}{2}s_{2k}^2 - 2\gamma^2 - 2\gamma$, then by (5.28), $2\gamma^2 + 2\gamma \geq \frac{3}{2}s_{2k}^2$. Therefore we have

$$2\gamma_1^2 + 2\gamma_1 \geq \frac{3}{2}s_{2k}^2. \tag{5.30}$$

On the other hand, as $\min_{q>0} g_{\gamma_1}(q) = g_{\gamma_0}(r) < 2s_{2k}^2$, one concludes from (5.28) that

$$2\gamma_1^2 + 2\gamma_1 + \frac{1}{2}s_{2k}^2 < 2s_{2k}^2. \tag{5.31}$$

Then (5.30) and (5.31) lead to the desired contradiction, and Lemma 5.3.4 is proved. $\square$

We now return to the proof of Proposition 5.3.2. In view of Lemma 5.3.3 and formula (5.24) one concludes that the zeroes of $a^2 - 4b^2$ and $f(\sqrt{r})^2 - 4s_{2k}^2$ as well as the signs of these two expressions coincide.

(i) Assume that $f(\sqrt{r})^2 - 4s_{2k}^2 \geq 0$. We then conclude that $a^2 - 4b^2 \geq 0$. Further, by Lemma 5.3.4, $a \geq 0$. In view of (5.19), the eigenvalues of A are then given by $\lambda_{1,2} = \pm(\mu_+)^{\frac{1}{2}}$, $\lambda_{3,4} = \pm(\mu_-)^{\frac{1}{2}}$, where $\mu_\pm = -\frac{a}{2} \pm \frac{1}{2}\sqrt{a^2 - 4b^2} \in \mathbb{R}$. As $a \geq 0$ it then follows that $\mu_\pm \leq 0$. Hence $(\lambda_i)_{1 \leq i \leq 4}$ are purely imaginary, i.e. $\pm(0, 0, r_1, 0, 0, r_2)$ are (possibly degenerate) elliptic fixed points of X_γ.

(ii) Assume that $f(\sqrt{r})^2 - 4s_{2k}^2 < 0$. Then by (5.23),

$$\mu_\pm = -\frac{a}{2} \pm \frac{i}{2}|d_{1,\gamma}r_1 - d_{2,\gamma}r_2|\sqrt{4r_1r_2 - (d_{1,\gamma}r_1 + d_{2,\gamma}r_2)^2}.$$

By Lemma 5.3.3 it then follows that Im $\mu_\pm \neq 0$, and we conclude that $\lambda_{1,2} = \pm(\mu_+)^{\frac{1}{2}}$, $\lambda_{3,4} = \pm(\overline{\mu}_+)^{\frac{1}{2}}$. In particular, two eigenvalues have a positive real part and the other two a negative real part. Hence $\pm(0, 0, r_1, 0, 0, r_2)$ are both hyperbolic fixed points of X_γ, and the corresponding stable and unstable manifolds have each dimension two. $\square$

5.3. FOLIATION OF $\mathcal{P}_K$ FOR $0 < K < \frac{N}{4}$

Having thus proved the hyperbolicity criterion (5.17), we turn to the geometry of $\mathcal{M}_\gamma^{-1}\{(\varepsilon(r_1 - r_2), 0)\}$. For the remaining statements we separately treat for any given $1 \leq k < \frac{N}{4}$ three subsets of the domain of the parameters γ and r. The results for these three cases are stated in detail in Propositions 5.3.5, 5.3.6, and 5.3.7 below. We emphasize that this part of the proof of Theorem 1.4.1 is almost exclusively due to T. Kappeler.

Note that the inverse image $\mathcal{M}_\gamma^{-1}\{(\varepsilon(r_1-r_2), 0)\}$ is invariant under the action of the vector field Y. The orbits of this action can be easily described,

$$\{(R_{-\phi}(M_1, J_1), L_1, R_\phi(M_2, J_2), L_2) \big| |\phi| \leq \pi\}, \quad (5.32)$$

where $R_\phi(u, v)$ denotes the image of $(u, v) \in \mathbb{R}^2$ under the rotation R_ϕ by the angle ϕ in counterclockwise orientation; in particular, these orbits are periodic. Hence given L_1 and L_2 with $|L_i| < r_i$ for $i = 1, 2$ there exists a unique point $(\hat{M}_i, \hat{J}_i, L_i)_{1 \leq i \leq 2}$ on such an orbit satisfying

$$(\hat{M}_1, \hat{J}_1) = \left(\sqrt{r_1^2 - L_1^2}, 0\right), \quad (\hat{M}_2, \hat{J}_2) = \sqrt{r_2^2 - L_2^2}\,(\cos\alpha, \sin\alpha) \quad (5.33)$$

for some $0 \leq \alpha < 2\pi$. We denote the corresponding Y-orbit by $\mathcal{L}(L_1, L_2, \alpha)$, i.e.

$$\mathcal{L}(L_1, L_2, \alpha) = \left\{\left(R_{-\varphi}\left(\sqrt{r_1^2 - L_1^2}, 0\right), L_1, R_{\alpha+\varphi}\left(\sqrt{r_2^2 - L_2^2}, 0\right), L_2\right) \Big| |\phi| \leq \pi\right\}. \quad (5.34)$$

As G and K_γ commute, K_γ is invariant along any orbit of the vector field Y, and we conclude from (5.13) and (5.33) that

$$K_\gamma((M_i, J_i, L_i)_{1 \leq i \leq 2}) = \frac{1}{2}\sum_{i=1}^{2} d_{i,\gamma}(r_i^2 - L_i^2) + \sqrt{(r_1^2 - L_1^2)(r_2^2 - L_2^2)}\cos\alpha. \quad (5.35)$$

Let us now determine $\mathcal{M}_\gamma^{-1}\{(\varepsilon(r_1-r_2), 0)\}$ for $\varepsilon = 1$ and $r \leq 1$. (The case where $r > 1$ and/or $\varepsilon = -1$ is treated similarly.) Let $(M_i, J_i, L_i)_{1 \leq i \leq 2}$ be an element of $\mathcal{M}_\gamma^{-1}\{(r_1 - r_2, 0)\} \setminus \{(0, 0, r_1, 0, 0, r_2)\}$. Then we have $r_2 - L_2 = r_1 - L_1$ and $K_\gamma((M_i, J_i, L_i)_{1 \leq i \leq 2}) = 0$. First note that $L_1 < r_1$, since if $L_1 = r_1$, then $L_2 = r_2$ and $(M_i, J_i) = (0, 0)$ for $i = 1, 2$, contradicting our assumption on the point considered. Hence in the expression (5.35) for K_γ we can factor out $r_1 - L_1$, and the equation $K_\gamma = 0$ reads

$$0 = d_{1,\gamma}(r_1 + L_1) + d_{2,\gamma}(r_2 + L_2) + 2\sqrt{(r_1 + L_1)(r_2 + L_2)}\cos\alpha. \quad (5.36)$$

Next let us consider the case where $L_1 = -r_1$. Then $(M_1, J_1, L_1) = -(0, 0, r_1)$ and (5.36) reads $(\gamma + c_k^2)(r_2 + L_2) = 0$. Hence either $L_2 = -r_2$ or $\gamma = -c_k^2$. In the case $L_2 = -r_2$ it follows from $r_2 - L_2 = r_1 - L_1 = 2r_1$ that $r_2 = r_1$ and $L_2 = L_1$. As a consequence $(M_2, J_2, L_2) = -(0, 0, r_1)$. On the other hand, if $\gamma = -c_k^2$ and $r_1 < r_2$, then

$$-r_2 < r_2 - 2r_1 = L_2 < r_2 \quad \text{and} \quad M_2^2 + J_2^2 = r_2^2 - L_2^2 = 4r_1(r_2 - r_1).$$

If $L_2^- \neq -r_1$ (and hence $L_2 \neq -r_2$ as $r_1 \leq r_2$), we set for $-r_1 < L_1 < r_1$

$$Q(L_1) := \begin{cases} \sqrt{\frac{r_1+L_1}{r_2+L_2(r_1)}} = \sqrt{\frac{r_1+L_1}{2r_2+L_1-r_1}} & \text{if } r_1 < r_2, \\ 1 & \text{if } r_1 = r_2. \end{cases} \quad (5.37)$$

Then $0 < Q(L_1) < \sqrt{r} \leq 1$ and for $r < 1$, Q is monotonically increasing in L_1 cn $-r_1 < L_1 < r_1$. After division by $\frac{1}{s_{2k}}\sqrt{(r_1+L_1)(r_2+L_2)}$ the equation (5.36) reads

$$(\gamma + s_k^2)Q(L_1) + (\gamma + c_k^2)\frac{1}{Q(L_1)} + 2s_{2k}\cos\alpha = 0. \quad (5.38)$$

To investigate the solutions of (5.38) we distinguish between the three cases $r=1$, $[\gamma + c_k^2 = 0$ and $r < 1]$, and $[\gamma + c_k^2 \neq 0$ and $r < 1]$.

We first treat the case $r = 1$. Then $Q \equiv 1$, and equation (5.38) takes the form

$$2\gamma + 1 = -2s_{2k}\cos\alpha, \quad (5.39)$$

which is in particular independent of L_1.

Proposition 5.3.5. *Let $\gamma \in \mathbb{R}$ be arbitrary and assume that $1 \leq k < \frac{N}{4}$ and $r = 1$. Then the following statements hold:*

(i) If $|2\gamma + 1| > 2s_{2k}$, then

$$\mathcal{M}_\gamma^{-1}\{(0,0)\} = \{\varepsilon(0,0,r_1,0,0,r_1) | \varepsilon = \pm\}, \quad (5.40)$$

and $\pm(0,0,r_1,0,0,r_1)$ are both elliptic fixed points of the vector field X_γ.

(ii) If $|2\gamma + 1| < 2s_{2k}$, then $\mathcal{M}_\gamma^{-1}\{(0,0)\} = \mathcal{N}_\alpha \cup \mathcal{N}_{-\alpha}$, where α is the unique angle satisfying $0 < \alpha < \pi$ and $2\gamma + 1 = -2s_{2k}\cos\alpha$, and where for any $-\pi \leq \beta \leq \pi$

$$\mathcal{N}_\beta = \bigcup_{\substack{|L_1| \leq r_1 \\ L_2 = L_1}} \mathcal{L}(L_1, L_2, \beta)$$

with $\mathcal{L}(L_1, L_2, \beta)$ given by (5.34). Both points, $\pm(0,0,r_1,0,0,r_1)$, are hyperbolic fixed points of X_γ, and their stable and unstable manifolds have each dimension two. The set $\mathcal{N}_\alpha \setminus \{\pm(0,0,r_1,0,0,r_1)\}$ consists of heteroclinic X_γ-orbits from $(0,0,r_1,0,0,r_1)$ to $-(0,0,r_1,0,0,r_1)$, whereas $\mathcal{N}_{-\alpha} \setminus \{\pm(0,0,r_1,0,0,r_1)\}$ consists of heteroclinic X_γ-orbits with opposite direction. Topologically, $\mathcal{M}_\gamma^{-1}\{(0,0)\}$ is a 2-dimensional torus, pinched at each of the two fixed points $\pm(0,0,r_1,0,0,r_1)$.

(iii) If $2\gamma + 1 = -2s_{2k}$, then $\alpha = 0$ and $\mathcal{M}_\gamma^{-1}\{(0,0)\} = \mathcal{N}_0$, whereas if $2\gamma + 1 = 2s_{2k}$, then $\alpha = \pi$ and $\mathcal{M}_\gamma^{-1}\{(0,0)\} = \mathcal{N}_\pi$. In both cases, $\pm(0,0,r_1,0,0,r_1)$ are elliptic fixed points of X_γ. On $\mathcal{N}_0 \cup \mathcal{N}_\pi$, any X_γ-orbit is periodic and coincides with the corresponding Y-orbit at least up to orientation.

5.3. FOLIATION OF $\mathcal{P}_K$ FOR $0 < K < \frac{N}{4}$

Proof. By Proposition 5.3.2, we only have to prove the claims on the geometry of $\mathcal{M}_\gamma^{-1}\{(0,0)\}$; the same remark holds for the proofs of Propositions 5.3.6 and 5.3.7.

(i) The identity (5.40) easily follows from (5.39): Under the hypothesis $|2\gamma + 1| > 2s_{2k}$, there is no solution to (5.39), which implies that the set $\mathcal{M}_\gamma^{-1}\{(0,0)\} \setminus \{\pm(0,0,r_1,0,0,r_1)\}$ is empty.

(ii) By the discussion preceding Proposition 5.3.5, in particular formula (5.34), it follows that the inverse image $\mathcal{M}_\gamma^{-1}\{(0,0)\}$ is given as claimed. To see that $\mathcal{N}_\alpha \setminus \{\pm(0,0,r_1,0,0,r_1)\}$ consists of heteroclinic orbits of the vector field X_γ, consider the third component $(X_\gamma)_3$ of X_γ (cf (5.15)). By (5.34), any element $(M_i, J_i, L_i)_{1 \leq i \leq 2}$ of $\mathcal{N}_\alpha$ is of the form

$$(M_1, J_1) = \sqrt{r_1^2 - L_1^2}(\cos\phi, -\sin\phi), \quad (M_2, J_2) = \sqrt{r_1^2 - L_1^2}(\cos(\alpha+\phi), \sin(\alpha+\phi)). \tag{5.41}$$

Thus

$$\begin{aligned}(X_\gamma)_3 &= -(M_1 J_2 + M_2 J_1) \\ &= -(r_1^2 - L_1^2)(\cos\phi \sin(\alpha+\phi) - \cos(\alpha+\phi)\sin\phi) \\ &= -(r_1^2 - L_1^2)\sin\alpha \end{aligned} \tag{5.42}$$

Hence $(X_\gamma)_3 < 0$ for any point in $\mathcal{N}_\alpha \setminus \{\pm(0,0,r_1,0,0,r_1)\}$. As the last component of X_γ coincides with the third one, it follows that any X_γ-orbit on the set $\mathcal{N}_\alpha \setminus \{\pm(0,0,r_1,0,0,r_1)\}$ originates from $(0,0,r_1,0,0,r_1)$ and ends in $-(0,0,r_1,0,0,r_1)$, i.e. these orbits are heteroclinic orbits as claimed. The orbits on $\mathcal{N}_{-\alpha}$ are analyzed in a similar way.

(iii) Clearly, if $2\gamma - 1 = -2s_{2k}$, one has $\mathcal{M}_\gamma^{-1}\{(0,0)\} = \mathcal{N}_0$, and one verifies in a straightforward way that the points $\pm(0,0,r_1,0,0,r_1)$ are elliptic fixed points of X_γ. According to (5.42), the third component $(X_\gamma)_3$ of X_γ vanishes identically on $\mathcal{N}_0$. Further, $2\gamma + 1 = -2s_{2k}$ implies that $1 + d_{2,\gamma} = -1 - d_{1,\gamma}$. In view of (5.41) it then follows that

$$X_\gamma = (1 + d_{2,\gamma}) L_2 \cdot Y.$$

The claimed statements for the case $2\gamma + 1 = 2s_{2k}$ are proved in a similar fashion. $\square$

Next we consider the case where $r < 1$ and $\gamma + c_k^2 = 0$. Then $\gamma + s_k^2 = -c_{2k}$, and hence

$$d_{1,\gamma} = -\frac{c_{2k}}{s_{2k}}, \quad d_{2,\gamma} = 0.$$

Thus equation (5.38) takes the form

$$c_{2k} Q(L_1) = 2s_{2k} \cos\alpha; \tag{5.43}$$

note that by $1 \leq k < \frac{N}{4}$, $0 < c_{2k} < 1$; we extend $Q(L_1)$, defined by (5.37), continuously extended to the closed interval $[-r_1, r_1]$. We then have the estimate

$$0 \leq Q(L_1) \leq \sqrt{r}. \tag{5.44}$$

Proposition 5.3.6. *Assume that* $1 \leq k < \frac{N}{4}$, $0 < r < 1$, *and* $\gamma + c_k^2 = 0$. *Then the following statements hold:*

(i) *If* $\sqrt{r} > 2s_{2k}/c_{2k}$, *then the connected component of* $\mathcal{M}_\gamma^{-1}\{(r_1 - r_2, 0)\}$ *containing the critical point* $(0, 0, r_1, 0, 0, r_2)$ *consists of this point alone. It is an elliptic fixed point of* X_γ.

(ii) *If* $\sqrt{r} \leq 2s_{2k}/c_{2k}$, *then*

$$\mathcal{M}_\gamma^{-1}\{(r_1 - r_2, 0)\} = \bigcup_{\substack{|L_1| \leq r_1 \\ L_2 = L_1 + r_2 - r_1}} \mathcal{L}(L_1, L_2, \alpha_{L_1}) \cup \mathcal{L}(L_1, L_2, -\alpha_{L_1})$$

where for any $|L_1| \leq r_1$, α_{L_1} *is the unique angle satisfying*

$$c_{2k} Q(L_1) = 2s_{2k} \cos \alpha_{L_1} \quad \text{and} \quad 0 \leq \alpha_{L_1} \leq \frac{\pi}{2}.$$

Furthermore, the connected component of $\mathcal{M}_\gamma^{-1}\{(r_1 - r_2, 0)\}$ *containing* $(0, 0, r_1, 0, 0, r_2)$ *consists of homoclinic* X_γ-*orbits which originate and end in* $(0, 0, r_1, 0, 0, r_2)$. *Topologically, it is a 2-dimensional torus, pinched at* $(0, 0, r_1, 0, 0, r_2)$.

If $\sqrt{r} < 2s_{2k}/c_{2k}$, *then* $(0, 0, r_1, 0, 0, r_2)$ *is a hyperbolic fixed point of* X_γ *and its stable and unstable manifold have each dimension two. If* $\sqrt{r} = 2s_{2k}/c_{2k}$, $(0, 0, r_1, 0, 0, r_2)$ *is an elliptic fixed point of* X_γ.

Proof. (i) In view of equation (5.43), the estimate (5.44), and the discussion of the case $L_1 = \pm r_1$, item (i) follows easily.

(ii) Again by the discussion preceding Proposition 5.3.5, it follows that the inverse image $\mathcal{M}_\gamma^{-1}\{(r_1 - r_2, 0)\}$ is given as claimed. Next consider a point $(M_i, J_i, L_i)_{1 \leq i \leq 2}$ in $\mathcal{M}_\gamma^{-1}\{(r_1 - r_2, 0)\}$ with

$$(M_1, J_1) = \sqrt{r_1^2 - L_1^2} \, (\cos \phi, -\sin \phi),$$
$$(M_2, J_2) = \sqrt{r_2^2 - L_2^2} \, (\cos(\alpha_{L_1} + \phi), \sin(\alpha_{L_1} + \phi))$$

where $|L_1| < r_1$, i.e. an element of $\mathcal{L}(L_1, L_2, \alpha_{L_1})$. Then, by (5.42), the third component of X_γ is given by

$$(X_\gamma)_3 = -\sqrt{r_1^2 - L_1^2} \sqrt{r_2^2 - L_2^2} \sin \alpha_{L_1}.$$

Note that by the assumptions $\sqrt{r} \leq \frac{2s_{2k}}{c_{2k}}$ and $|L_1| < r_1$ (implying $Q(L_1) < \sqrt{r}$), it follows that $\alpha_{L_1} > 0$. Hence $(X_\gamma)_3 = (X_\gamma)_6 < 0$. It follows that the X_γ-orbit passing through such a point originates at $(0, 0, r_1, 0, 0, r_2)$ and then reaches a point of the form $(0, 0, -r_1, M_2, J_2, L_2)$ with

$$L_2 = r_2 - 2r_1 > -r_2, \quad (M_2, J_2) = \sqrt{r_2^2 - L_2^2} (\cos(\pi + \tilde{\phi}), \sin(\pi + \tilde{\phi})). \quad (5.45)$$

5.3. FOLIATION OF $\mathcal{P}_K$ FOR $0 < K < \frac{N}{4}$

for some $\tilde{\phi} \in \mathbb{R}/2\pi\mathbb{Z}$. At this point the vector field X_γ is given by (recall (5.15) and that $d_{2,\gamma} = 0$)
$$(-r_1 J_2, -r_1 M_2, 0, 0, 0, 0).$$
Note that this vector does not vanish as $M_2^2 + J_2^2 = r_2^2 - L_2^2 > 0$. Similarly, at a point $(M_i, J_i, L_i)_{1 \leq i \leq 2}$ in $\mathcal{M}_\gamma^{-1}\{(r_1 - r_2, 0)\}$ satisfying
$$(M_1, J_1) = \sqrt{r_1^2 - L_1^2}\,(\cos\phi, -\sin\phi),$$
$$(M_2, J_2) = \sqrt{r_2^2 - L_2^2}\,(\cos(-\alpha_{L_1} + \phi), \sin(-\alpha_{L_1} + \phi))$$
and $|L_1| < r_1$, i.e. an element of $\mathcal{L}(L_1, L_2, -\alpha_{L_1})$, one has
$$(X_\gamma)_3 = \sqrt{r_1^2 - L_1^2}\,\sqrt{r_2^2 - L_2^2}\,\sin\alpha_{L_1}.$$
Hence $(X_\gamma)_3 = (X_\gamma)_6 > 0$. It follows that the X_γ-orbit passing through such a point passes through a point of the form $(0, 0, -r_1, M_2, J_2, L_2)$ with (M_2, J_2, L_2) as in (5.45) and ends up at $(0, 0, r_1, 0, 0, r_2)$. We then conclude that the connected component of $\mathcal{M}_\gamma^{-1}\{(r_1 - r_2, 0)\}$ containing $(0, 0, r_1, 0, 0, r_2)$ consists of homoclinic X_γ-orbits originating and ending at $(0, 0, r_1, 0, 0, r_2)$. $\square$

Finally we treat the case $r < 1$ and $\gamma + c_k^2 \neq 0$. Again denote by $Q(L_1)$ the function defined by (5.37), extended continuously to the closed interval $[-r_1, r_1]$. Note that in this case, the function f, introduced in (5.18), satisfies $\lim_{q \searrow 0} |f(q)| = \infty$. Then (5.38) takes the form
$$f(Q(L_1)) + 2s_{2k} \cos\alpha = 0. \tag{5.46}$$

Proposition 5.3.7. *Assume that $1 \leq k < \frac{N}{4}$, $0 < r < 1$, and $\gamma + c_k^2 \neq 0$. Then the following statements hold:*

(i) *If $|f(\sqrt{r})| \geq 2s_{2k}$, then the connected component of $\mathcal{M}_\gamma^{-1}\{(r_1 - r_2, 0)\}$ containing the critical point $(0, 0, r_1, 0, 0, r_2)$ consists of this point alone. It is an elliptic fixed point of X_γ.*

(ii) *If $|f(\sqrt{r})| < 2s_{2k}$, then there exists $-r_1 < l_{\gamma,r} < r_1$ so that the connected component of $\mathcal{M}_\gamma^{-1}\{(r_1 - r_2, 0)\}$ containing $(0, 0, r_1, 0, 0, r_2)$ is given by*
$$\bigcup_{\substack{l_{\gamma,r} \leq L_1 \leq r_1 \\ L_2 = L_1 + r_2 - r_1}} \mathcal{L}(L_1, L_2, \alpha_{L_1}) \cup \mathcal{L}(L_1, L_2, -\alpha_{L_1})$$
where for any $l_{\gamma,r} \leq L_1 \leq r_1$, α_{L_1} is the unique angle satisfying $0 \leq \alpha_{L_1} \leq \pi$ and
$$f(Q(L_1)) = -2s_{2k} \cos(\alpha_{L_1}).$$
The point $(0, 0, r_1, 0, 0, r_2)$ is a hyperbolic fixed point of X_γ and its stable and unstable manifold each have dimension two. The connected component of $\mathcal{M}_\gamma^{-1}\{(r_1 - r_2, 0)\}$ containing $(0, 0, r_1, 0, 0, r_2)$ consists of homoclinic X_γ-orbits which originate and end in $(0, 0, r_1, 0, 0, r_2)$. Topologically, it is a 2-dimensional torus, pinched at $(0, 0, r_1, 0, 0, r_2)$.

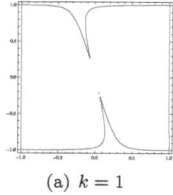 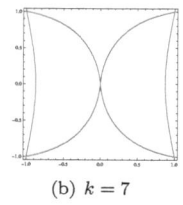 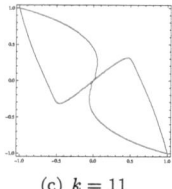

(a) $k = 1$ (b) $k = 7$ (c) $k = 11$

Figure 5.1: Sets of solutions (l_1, l_2) of (5.55) for $N = 48$, $r = 1$, $\gamma = -1.35$

Proof. (i) We have already seen that under the given assumption $\mathcal{M}_\gamma^{-1}\{(r_1 - r_2, 0)\} \setminus \{(0, 0, r_1, 0, 0, r_2)\}$ consists of the set of points $(M_i, J_i, L_i)_{1 \leq i \leq 2}$ satisfying $L_2 = L_1 + r_2 - r_1$ and (5.38). Note that equation (5.46) admits a solution α for $Q = \sqrt{r}$ if and only if $|f(\sqrt{r})| \leq 2s_{2k}$. In the case $|f(\sqrt{r})| > 2s_{2k}$ it follows immediately that $\mathcal{M}_\gamma^{-1}\{(r_1 - r_2, 0)\} = \{(0, 0, r_1, 0, 0, r_2)\}$. If $|f(\sqrt{r})| = 2s_{2k}$, then an analysis of the graph of f near $(\sqrt{r}, f(\sqrt{r}))$ leads to the claimed result.

(ii) As $\lim_{q \searrow 0} |f(q)| = \infty$, it follows that there exists $-r_1 < l_{\gamma,r} < r_1$ so that the interval $[l_{\gamma,r}, r_1]$ is a connected component of $(f \circ Q)^{-1}([-2s_{2k}, 2s_{2k}])$. Hence for any $l_{\gamma,r} \leq L_1 \leq r_1$ there exists a unique angle $0 \leq \alpha_{L_1} \leq \pi$ so that (5.46) is satisfied for $\alpha = \alpha_{L_1}$. Again by recalling (5.34), the connected component of the preimage $\mathcal{M}_\gamma^{-1}\{(r_1 - r_2, 0)\}$ containing the point $(0, 0, r_1, 0, 0, r_2)$ is then given as claimed. One then can argue as in the proof of item (ii) of Proposition 5.3.6 to show the remaining claims. □

Proof of Theorem 1.4.1. As explained above, Theorem 1.4.1 follows directly from Proposition 5.3.2 and Propositions 5.3.5 - 5.3.7. □

It remains to study the critical points of $\mathcal{M}_\gamma$ with rank one, i.e. points of $(\mathbb{S}_{r_1}^2 \times \mathbb{S}_{r_2}^2) \setminus \{\pm(0, 0, r_1, 0, 0, \pm r_2)\}$ where the vector fields Y and X_γ are collinear, but do not both vanish. In view of the formulas (5.15) for Y and X_γ, points $(M_i, J_i, L_i) \in S_{r_i}^2$, $i = 1, 2$, of this type have the property that any 2×2-submatrix of the 2×4-matrix formed by Y and X_γ is singular. This leads to the following system of equations:

$$M_1 J_2 + M_2 J_1 = 0, \qquad (5.47)$$

$$J_1^2 L_2 + L_1 J_2^2 - J_1 J_2 (d_{1,\gamma} L_1 + d_{2,\gamma} L_2) = 0, \qquad (5.48)$$

$$M_1^2 L_2 + L_1 M_2^2 + M_1 M_2 (d_{1,\gamma} L_1 + d_{2,\gamma} L_2) = 0, \qquad (5.49)$$

$$M_1 J_1 L_2 - L_1 M_2 J_2 + J_1 M_2 (d_{1,\gamma} L_1 + d_{2,\gamma} L_2) = 0. \qquad (5.50)$$

5.3. FOLIATION OF $\mathcal{P}_K$ FOR $0 < K < \frac{N}{4}$

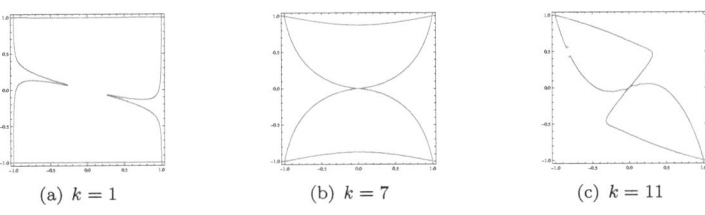

Figure 5.2: Sets of solutions (l_1, l_2) of (5.55) for $N = 48$, $r = 1$, $\gamma = 0.35$

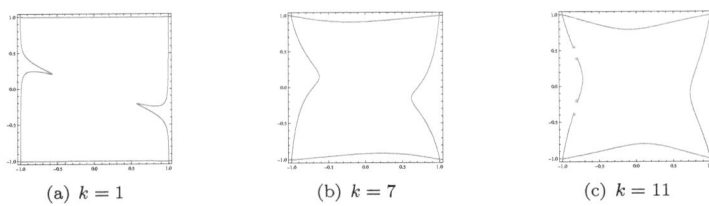

Figure 5.3: Sets of solutions (l_1, l_2) of (5.55) for $N = 48$, $r = 1$, $\gamma = 0.6$

Proof of Theorem 1.4.2. First assume that $L_1 \in \{\pm r_1\}$. Then $J_1 = M_1 = 0$. Hence (5.47) is automatically satisfied and equations (5.48) and (5.49) read $J_2 = 0$ and $M_2 = 0$, respectively. As a consequence, $(M_1, J_1, L_1) = (0, 0, \pm r_1)$ and $(M_2, J_2, L_2) = (0, 0, \pm r_2)$, which are the critical points of $\mathcal{M}_\gamma$ with rank zero, contradicting the assumption rank $d\mathcal{M}_\gamma = 1$. We thus may assume that $|L_1| < r_1$. Then $(M_1, J_1) \neq (0, 0)$. Hence the first equation (5.47) implies that there exists $\lambda \in \mathbb{R}$ such that

$$(M_2, J_2) = \lambda(M_1, -J_1). \tag{5.51}$$

The conditions $(M_i, J_i, L_i) \in \mathbb{S}^2_{r_i}$, $i = 1, 2$, then imply that λ satisfies

$$\lambda^2 = \frac{r_2^2 - L_2^2}{r_1^2 - L_1^2}. \tag{5.52}$$

Substituting (5.51) into (5.48)-(5.50) one sees, again using $(M_1, J_1) \neq (0, 0)$, that (5.48)-(5.50) is equivalent to the condition

$$L_2 + \lambda^2 L_1 + \lambda(d_{1,\gamma} L_1 + d_{2,\gamma} L_2) = 0, \tag{5.53}$$

or, taking squares, $(L_2 + \lambda^2 L_1)^2 - \lambda^2(d_{1,\gamma} L_1 + d_{2,\gamma} L_2)^2 = 0$. Using (5.52), the latter equation reads

$$(r_1^2 - L_1^2)^2 L_2^2 + (r_2^2 - L_2^2)^2 L_1^2 + 2(r_1^2 - L_1^2)(r_2^2 - L_2^2)(2L_1 L_2 - (d_{1,\gamma} L_1 + d_{2,\gamma} L_2)^2) = 0. \tag{5.54}$$

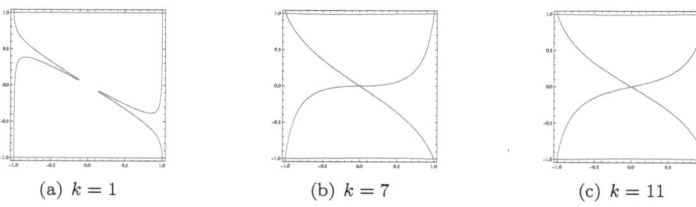

Figure 5.4: Sets of solutions (l_1, l_2) of (5.55) for $N = 48$, $r = 0.3$, $\gamma = -2$

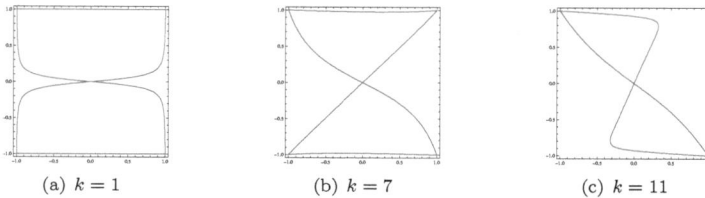

Figure 5.5: Sets of solutions (l_1, l_2) of (5.55) for $N = 48$, $r = 0.3$, $\gamma = -1$

After dividing by $r_1^2 r_2^4$ one gets, using the bifurcation parameter r and the normalized variables $l_i := \frac{L_i}{r_i} \in (0,1)$ ($i = 1, 2$),

$$r^2(1-l_1^2)^2 l_2^2 + (1-l_2^2)^2 l_1^2 + 2r(1-l_1^2)(1-l_2^2)(2l_1 l_2 - (\sqrt{r} d_{1,\gamma} l_1 + \frac{1}{\sqrt{r}} d_{2,\gamma} l_2)^2) = 0. \quad (5.55)$$

Note that for given r and $0 < l_1 < 1$, the left hand side of (5.55) is a polynomial in l_2 of degree four, i.e. $(l_1^2 + 2d_{2,\gamma}^2(1-l_1^2))l_2^4 + O(l_2^3)$. For any of the four possible solutions l_2 of (5.55), we obtain a solution L_2 of (5.54) and two possible solutions λ of (5.52), which, for given (M_1, J_1), determines (M_2, J_2) through (5.51).

Summarizing, we have shown that for any given point in $\mathbb{S}_{r_1}^2 \setminus \{(0,0,\pm r_1)\}$, there exist at most eight points in $\mathbb{S}_{r_2}^2 \setminus \{(0,0,\pm r_2)\}$ such that Y and X_γ are collinear. Indeed for any $(M_1, J_1, L_1) \in \mathbb{S}_{r_1}^2 \setminus \{(0,0,\pm r_1)\}$, a solution $(M_2, J_2, L_2) \in \mathbb{S}_{r_2}^2 \setminus \{(0,0,\pm r_2)\}$ of (5.47)-(5.50) is given by $(M_2, J_2) = \lambda(M_1, J_1)$ and $L_2 = r_2 l_2$ with λ and l_2 satisfying (5.52) and (5.55), respectively. As a consequence the set of solutions of (5.47)-(5.50) is an algebraic subset of $\mathbb{S}_{r_1}^2 \times \mathbb{S}_{r_2}^2$ of dimension at most two. □

In order to analyze these critical points of $\mathcal{M}_\gamma$ with rank one, we perform another symplectic reduction. First we pass to the orbit space of the flow of Y on $\mathbb{S}_{r_1}^2 \times \mathbb{S}_{r_2}^2$ and then to the level sets of G.

In view of (5.32), the Y-flow is an $\mathbb{S}^1$-action on $\mathbb{S}_{r_1}^2 \times \mathbb{S}_{r_2}^2$. Note that besides

5.3. FOLIATION OF $\mathcal{P}_K$ FOR $0 < K < \frac{N}{4}$

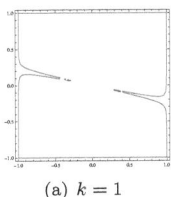
(a) $k = 1$

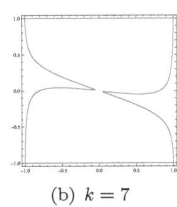
(b) $k = 7$

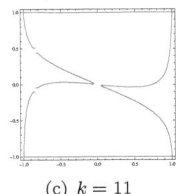
(c) $k = 11$

Figure 5.6: Sets of solutions (l_1, l_2) of (5.55) for $N = 48$, $r = 0.3$, $\gamma = 3$

L_1 and L_2, the quantities σ and τ,

$$\sigma := M_1 M_2 - J_1 J_2, \quad \tau := M_1 J_2 + M_2 J_1,$$

are invariant under this $\mathbb{S}^1$-action. They are related by

$$\sigma^2 + \tau^2 = \prod_{i=1}^{2}(r_i^2 - L_i^2). \tag{5.56}$$

Define for $\dot{\mathbb{S}}_{r_i}^2 := \mathbb{S}_{r_i}^2 \setminus \{(0, 0, \pm r_i)\}$, $i = 1, 2$,

$$\begin{aligned}
\mathcal{F}^{(3)} : \quad & \dot{\mathbb{S}}_{r_1}^2 \times \dot{\mathbb{S}}_{r_2}^2 \quad \to \quad \mathbb{R}^4 \\
& (M_1, J_1, L_1, M_2, J_2, L_2) \mapsto (L_1, L_2, \sigma, \tau),
\end{aligned}$$

and let $\mathcal{O}_r$ denote the image of $\mathcal{F}^{(3)}$. For any element $(L_1, L_2, \sigma, \tau) \in \mathcal{O}_r$ we have

$$\sigma^2 + \tau^2 = \prod_{i=1}^{2}(r_i^2 - L_i^2) \quad \text{and} \quad |L_i| < r_i \ (i = 1, 2). \tag{5.57}$$

The fibers of $\mathcal{F}^{(3)}$ are the orbits of the Y-action on $\dot{\mathbb{S}}_{r_1}^2 \times \dot{\mathbb{S}}_{r_2}^2$, i.e. $\mathcal{O}_r$ coincides with the orbit space of the Y-action on $\dot{\mathbb{S}}_{r_1}^2 \times \dot{\mathbb{S}}_{r_2}^2$. Consequently, any function on $\dot{\mathbb{S}}_{r_1}^2 \times \dot{\mathbb{S}}_{r_2}^2$ which Poisson commutes with G factors through $\mathcal{O}_r$.

In particular, K_γ and G factor through $\mathcal{O}_r$. In fact, K_γ and G, when expressed in the variables L_1, L_2, σ, τ, are polynomials, given by (cf. (5.12) and (5.13))

$$K_\gamma = \sum_{i=1}^{2} \frac{1}{2} d_{i,\gamma}(r_i^2 - L_i^2) + \sigma, \tag{5.58}$$

$$G = L_1 - L_2. \tag{5.59}$$

By reducing the system (G, K_γ) by the Y-action we obtain a family of integrable systems with one degree of freedom parametrized by the value c of G. Denote by $X_{\gamma,c}$ the Hamiltonian vector field induced by K_γ on the level set of G to the value c. As above, the fixed points of $X_{\gamma,c}$ can then be characterized in terms of the bifurcation parameters γ, r, and k.

Note that by (5.47), the rank-1-points of the reduced moment map $\mathcal{M}_\gamma$ satisfy $\tau = 0$, and by (5.56), $\sigma^2 = (r_2^2 - L_2^2)(r_1^2 - L_1^2)$. Hence the image of the set of the rank-1-points by $\mathcal{F}^{(3)}$ is an algebraic subset of $\mathcal{O}_r$ of dimension at most one - see (5.54).

By (5.56), σ and τ are located on a circle of radius $\sqrt{(r_1^2 - L_1^2)(r_2^2 - L_2^2)}$,

$$(\sigma, \tau) = \sqrt{(r_1^2 - L_1^2)(r_2^2 - L_2^2)} \, (\cos\phi, \sin\phi), \tag{5.60}$$

where $\phi \in \mathbb{R}/2\pi\mathbb{Z}$. The phase spaces, reduced by the Y-action, are then obtained by taking subsets of $\mathcal{O}_r$ corresponding to level sets of G, i.e. by replacing L_2 by $L_1 - c$, where c is the value of G. The restriction $K_{\gamma,c}$ of K_γ to the reduced phase space corresponding to the value c of G is then given by

$$K_{\gamma,c}(L_1, \phi) = \frac{1}{2} \left(d_{1,\gamma}(r_1^2 - L_1^2) + d_{2,\gamma}(r_2^2 - (L_1 - c)^2) \right)$$
$$+ \sqrt{(r_1^2 - L_1^2)(r_2^2 - (L_1 - c)^2)} \cos\phi \tag{5.61}$$

with $L_1 \in ((-r_1, r_1) \cap (c - r_2, c + r_2))$ and $\phi \in \mathbb{R}/2\pi\mathbb{Z}$.

Again by using the general scheme (5.9), the reduced Hamiltonian vector field induced by $K_{\gamma,c}$ is given by

$$X_{\gamma,c}(L_1, \phi) = \frac{d}{dt}\begin{pmatrix} L_1 \\ \phi \end{pmatrix} = \{L_1, \phi\} \begin{pmatrix} \partial K_{\gamma,c}/\partial\phi \\ -\partial K_{\gamma,c}/\partial L_1 \end{pmatrix}. \tag{5.62}$$

The Poisson bracket $\{L_1, \phi\}$ can be computed to be

$$\{L_1, \phi\} = \left\{L_1, \arctan\frac{\tau}{\sigma}\right\} = \frac{1}{1 + (\tau/\sigma)^2}\left\{L_1, \frac{\tau}{\sigma}\right\}$$
$$= \frac{1}{1 + (\tau/\sigma)^2} \cdot \frac{\sigma\{L_1, \tau\} - \tau\{L_1, \sigma\}}{\sigma^2} = \frac{1}{1 + (\tau/\sigma)^2} \cdot \frac{\sigma^2 + \tau^2}{\sigma^2} = (5.63)$$

since by the commutation relations among the variables L_1, M_1, J_1 and Lemma 3.1.8, $\{L_1, \tau\} = \sigma$ and $\{L_1, \sigma\} = -\tau$. Furthermore, with $L_2 = L_1 - c$ (but treated as a dependent variable),

$$\frac{\partial K_{\gamma,c}}{\partial \phi} = -\sqrt{(r_1^2 - L_1^2)(r_2^2 - L_2^2)} \sin\phi,$$

$$\frac{\partial K_{\gamma,c}}{\partial L_1} = -(d_{1,\gamma}L_1 + d_{2,\gamma}L_2) - \frac{L_1(r_2^2 - L_2^2) + L_2(r_1^2 - L_1^2)}{\sqrt{(r_1^2 - L_1^2)(r_2^2 - L_2^2)}} \cos\phi.$$

Hence, by (5.63), (5.62) reads

$$X_{\gamma,c}(L_1, \phi) = \begin{pmatrix} -\sqrt{(r_1^2 - L_1^2)(r_2^2 - L_2^2)} \sin\phi \\ (d_{1,\gamma}L_1 + d_{2,\gamma}L_2) + \frac{L_1(r_2^2 - L_2^2) + L_2(r_1^2 - L_1^2)}{\sqrt{(r_1^2 - L_1^2)(r_2^2 - L_2^2)}} \cos\phi \end{pmatrix}. \tag{5.64}$$

It follows from (5.64) that the fixed points of the vector field $X_{\gamma,c}$ with $|L_1| < r_1$ are given by (L_1, ϕ) satisfying

$$\phi \in \pi\mathbb{Z} \tag{5.65}$$

5.3. FOLIATION OF $\mathcal{P}_K$ FOR $0 < K < \frac{N}{4}$

and

$$(d_{1,\gamma} L_1 + d_{2,\gamma} L_2) + \frac{L_1(r_2^2 - L_2^2) + L_2(r_1^2 - L_1^2)}{\sqrt{(r_1^2 - L_1^2)(r_2^2 - L_2^2)}} \cos \phi = 0. \tag{5.66}$$

Note that (5.65) and the square of (5.66) are equivalent to the system of equations (5.47)-(5.55) derived above.

In order to determine the type of the fixed points (L_1, ϕ), i.e. points satisfying (5.65)-(5.66) for a given value c of G, one computes the Jacobian of $X_{\gamma,c}$ at these points, which we write as $H \equiv H_{\gamma,c}(L_1, \phi) = \begin{pmatrix} h_{11} & h_{12} \\ h_{21} & h_{22} \end{pmatrix}$. Note that by (5.65), we have $h_{11} = 0$ and $h_{22} = 0$ at such points, and thus $\det(H) = -h_{12} h_{21}$. Hence such a fixed point is an elliptic or hyperbolic fixed point of $X_{\gamma,c}$ if $h_{12} h_{21}$ is positive or negative, respectively. We investigate the signs of h_{12} and h_{21} separately.

First note that

$$h_{12} = -\sqrt{(r_1^2 - L_1^2)(r_2^2 - L_2^2)} \cos \phi,$$

hence, since $\phi \in \pi \mathbb{Z}$, $\text{sign}(h_{12}) = -(-1)^{\frac{\phi}{\pi}}$. Next, we compute

$$h_{21} = \left(-(d_{1,\gamma} + d_{2,\gamma}) \pm \frac{\partial}{\partial L_1} \left(\frac{L_1(r_2^2 - (L_1 - c)^2) + L_2(r_1^2 - L_1^2)}{\sqrt{(r_1^2 - L_1^2)(r_2^2 - (L_1 - c)^2)}} \right) \right),$$

where the sign in front of the derivative is again equal to $(-1)^{\frac{\phi}{\pi}}$. Another (lengthy) computation shows that the latter derivative is equal to the nonnegative expression

$$\frac{1}{\sqrt{(1-l_1^2)(1-l_2^2)}} \cdot \left(\sqrt{r \cdot \frac{1-l_1^2}{1-l_2^2}} - \left(\sqrt{r \cdot \frac{1-l_1^2}{1-l_2^2}} \right)^{-1} \right)^2,$$

again with $l_i := L_i/r_i$ $(i = 1, 2)$. Since $d_{1,\gamma} + d_{2,\gamma} = \frac{2\gamma+1}{s_{2k}}$, the classification reduces to investigating for solutions of (5.55) the sign of

$$\frac{2\gamma+1}{s_{2k}} \pm \frac{1}{\sqrt{(1-l_1^2)(1-l_2^2)}} \cdot \left(\sqrt{r \cdot \frac{1-l_1^2}{1-l_2^2}} - \left(\sqrt{r \cdot \frac{1-l_1^2}{1-l_2^2}} \right)^{-1} \right)^2.$$

Chapter 6

Discussion and Outlook

The fact that although at first glance, one-dimensional FPU chains appear to be a rather simple system, they exhibit at the same time such a rich dynamics, leads to several conclusions, some mathematical ones concerning perturbation theory in general and some physical or "general" ones concerning the FPU paradox and its explanation. Let us first turn to the former issue.

First of all, it is the "simplicity" of the system under consideration which makes it possible to carry through all necessary computations in order to obtain normal forms of an order high enough to check the hypotheses of the KAM and Nekhoroshev theorems. Especially in the case of the latter theorem, it is precisely the difficulty of rigorously checking its assumptions which seems to be the reason that the number of systems, to which the Nekhoroshev theorem has been explicitly applied, is rather small. In particular, it seems that up to now, there have not been many thorough discussions of Nekhoroshev's original criteria for "steepness" at a given example. Since these criteria apparently are considerably weaker than convexity or even (directional) quasi-convexity, such investigations could greatly extend the class of systems to which Nekhoroshev's estimates apply. Similar things can be said on Rüssmanns higher order nondegeneracy conditions, which are a weaker version of Kolmogorov's original nondegeneracy conditions (these weaker conditions have apparently been thought of in order to deal with systems, primarily in celestial machanics, which are obviously not nondegenerate).

Even though FPU chains fail to meet the original nondegeneracy conditions only in some exceptional cases of the parameter values, we think that it would be worthwhile to try to check these higher order conditions in these exceptional cases, and it seems possible that this could be accomplished by simply pursuing further the approach of this work, namely explicitly computing the coefficients of the Birkhoff normal form up to higher and higher orders and then checking the appropriate conditions. Of course, it cannot be taken for granted that there are no resonances, i.e. obstructions to the transformation to Birkhoff normal forms up to higher orders, similarly to the case of even periodic chains, where there are fourth order resonances leading to a *resonant* normal form of order four. However, due to the fact that we know that the (full) periodic Toda lattice is an integrable system, we are confident that it should be possible to carry thorugh this procedure at least in some cases of the parameter values, namely those approximating the Toda lattice up to higher and higher orders.

Besides justifying the claims that the KAM and Nekhoroshev theorems can be applied to FPU chains in the cases of odd periodic and Dirichlet chains, we also have thoroughly investigated even periodic chains. First of all, it is a surprising fact that the truncated fourth order Hamiltonian of these chains turns out to be integrable for all parameter values. Even though we do not "explain" this integrability by abstract geometric or group-theoretic arguments, we think that this fact could also contribute to an explanation of the FPU paradox, depending on the results of the planned numerical implementation of our results. Moreover, the detailed analysis of the level sets of the associated moment map reveals an extremely rich geometric structure, which is also somewhat surprising in view of the fact that we have partitioned all the integrals of this integrable system into subsets of at most four integrals, i.e. the systems under investigation "live" in an a phase space with at most four degrees of freedom. Nevertheless, only after repeated reductions we have been able to properly classify the various critical points of the originally given moment map. Moreover, the bifurcation diagrams obtained at two different steps of this reduction process turn out to have a rich geometric structure themselves - we have tried to convey an impression of this structure by plotting some particularly interesting examples. However, there are a lot of questions concerning these bifurcations which we have not answered, for instance whether the "domains of hyperbolicity" are always connected, what their asymptotic behavior is in the case of the particle number tending to infinity, just to mention some of these questions. Similar questions could be posed for the set of critical points of the moment map of rank one - not only concerning their nature (hyperbolic or elliptic), but also their distribution in the plane of the two (normed) action variables, what the geometric properties of the set given by the solutions of the appropriate equations are. And again, we emphasize that all these questions arise from the analysis of a system in an eight-dimensional phase space. It is clear that similar systems on phase spaces of higher dimensions can become analytically intractable quite rapidly.

Returning to those FPU chains where we are able to compute Birkhoff normal forms of order four, the further computation of coefficients of an even higher order would also contribute to a preciser implementation of our transformation formulas - and this directly leads to the second issue to be discussed, namely the relevance of our results for the explanation of the FPU paradox. As already emphasized in the introduction, before having implemented our transformation formulas, we do not attempt to fully answer the question of the "explanatory" power of our work. However, we have rigorously justified the claim that in particular Dirichlet chains, i.e. those which were originally considered by Fermi, Pasta, and Ulam, can be considered as fifth-order perturbations of a nondegenerate integrable system, thereby confirming a conjecture which has been proposed repeatedly in the last forty years. Furthermore and more generally, it seems to be quite promising that it has turned out to be possible to approximate all three types of chains (including the even periodic ones) up to fourth order with integrable systems. But, again, it is explicitly not our attempt to work "against" the other approach towards a resolution of the FPU paradox, namely investigating the continuum limit of the discrete chain and explaining the behavior of the discrete chain by the "soliton-like" behavior of the continuous limit.

In this way, we arrive at another promising direction for future research, namely to tackle the problems arising at the "interface" of the discrete and the continuous models. A first step could be to analyze how our results behave in the limit of the number of particles tending to infinity. Of course, we do not claim that this has to be started "ab ovo" - many steps have already been undertaken in this direction.

Finally, the FPU problem is a "toy example" insofar as it has been constructed from idealized assumptions, let us only mention the assumptions of equal masses and only nearest-neighbor interaction. Whereas this makes all our very explicit investigations possible, it has the drawback that one should be extremely careful in drawing "realistic" conclusions from our results - recall also the remarks in the introduction on the epistemolgical issues of the FPU problem. Nevertheless, we think that rigorously understanding such a comparatively simple system can be of great help towards the explanation of more complex systems (also, and in particular, non-physical systems).

Appendix A

Nonresonance Lemma

For completeness, we provide a detailed proof of Lemma 3.1.4 in this appendix. This lemma and its proof are due to Beukers and Rink, see ([32], Appendix A). A very similar statement, from which Lemma 3.1.4 can be deduced as well, has been proved by Conway and Jones - see [15]. Recall from (3.2.) that $K_4 \setminus K_4^N \subseteq \mathbb{Z}^4$ denotes the subset of index quadruples (k_1, k_2, k_3, k_4) satisfying $1 \leq |k_i| \leq N-1$ ($1 \leq i \leq 4$) and $k_1 + k_2 + k_3 + k_4 \equiv 0 \bmod N$ so that there are no integers l, m with $\{l, m, -l, -m\} = \{k_1, k_2, k_3, k_4\}$. Further,

$$K_4^{res} := K_{res}^+ \cup K_{res}^- \subseteq K_4$$

where

$$K_{res}^\pm := \left\{ (k_1, k_2, k_3, k_4) \in K_4 \mid \exists l \in \mathbb{N} : 1 \leq l \leq \frac{N}{4} \text{ so that } \right.$$
$$\left. \{k_1, k_2, k_3, k_4\} = \{\pm l, \pm l \mp N, \frac{N}{2} \mp l, -\frac{N}{2} \mp l\} \right\}.$$

Note that $K_4^{res} = \emptyset$ if N is odd. We restate Lemma 3.1.4 as follows:

Lemma A.1. *Let $(k_1, k_2, k_3, k_4) \in K_4 \setminus K_4^N$. Then*

$$\sin \frac{k_1 \pi}{N} + \sin \frac{k_2 \pi}{N} + \sin \frac{k_3 \pi}{N} + \sin \frac{k_4 \pi}{N} = 0, \qquad (A.1)$$

if and only if $(k_1, k_2, k_3, k_4) \in K_4^{res}$.

Let us make a few preparations for the proof of Lemma A.1. By a straightforward computation one sees that the "if"-part of the claimed equivalence holds:

Lemma A.2. *For any $(k_1, k_2, k_3, k_4) \in K_4^{res}$, the identity (A.1) holds, i.e. one has $\sum_{i=1}^4 \sin \frac{k_i \pi}{N} = 0$.*

So it remains to prove the converse. In Lemma A.3 and Lemma A.4, we consider some special cases, before we treat the general case in Lemma A.5.

Lemma A.3. Let $(k_1, k_2, k_3, k_4) \in K_4 \setminus (K_4^N \cup K_4^{res})$. If there exist $l, m, n \in \mathbb{Z}$ such that

(i) $\{k_1, k_2, k_3, k_4\} = \{l, -l, m, n\}$, or

(ii) $\{k_1, k_2, k_3, k_4\} = \{l, N-l, m, n\}$ with $1 \leq l \leq N-1$, or

(iii) $\{k_1, k_2, k_3, k_4\} = \{l, -N-l, m, n\}$ with $-(N-1) \leq l \leq -1$,

then

$$\sum_{i=1}^{4} \sin \frac{k_i \pi}{N} \neq 0.$$

Proof. In the case (i), it follows that $m + n = N$ (and thus $1 \leq m, n \leq N - 1$) or $m + n = -N$ (and thus $-(N-1) \leq m, n \leq -1$). Hence for both of these possibilities, $\sin \frac{m\pi}{N}$ and $\sin \frac{n\pi}{N}$ have the same sign, and $\sum_{i=1}^{4} \sin \frac{k_i\pi}{N} = \sin \frac{m\pi}{N} + \sin \frac{n\pi}{N} \neq 0$. In the case (ii), by assumption, $m + n \equiv 0 \mod N$. The case $m + n = 0$ has already been treated under (i). If $m + n = N$, then $\sin \frac{k_i\pi}{N} > 0$ for any $1 \leq i \leq 4$. If $m + n = -N$, then $m < 0$, and $m \notin \{-l, -N+l\}$. Thus $n = -N - m < 0$ and therefore $\sum_{i=1}^{4} \sin \frac{k_i\pi}{N} = 2\sin\frac{l\pi}{N} - 2\sin\frac{(-m)\pi}{N} \neq 0$. The case (iii) is treated similarly to the case (ii). □

Lemma A.4. Assume that $(k_1, k_2, k_3, k_4) \in K_4 \setminus K_4^N$ satisfies

$$k_i + k_j \not\equiv 0 \mod N \quad \forall 1 \leq i, j \leq 4. \tag{A.2}$$

If there exist $l, n \in \{k_1, k_2, k_3, k_4\}$ with

$$\sin \frac{l\pi}{N} + \sin \frac{n\pi}{N} = 0, \tag{A.3}$$

then (A.1) implies that $(k_1, k_2, k_3, k_4) \in K_4^{res}$.

Proof. From the assumptions (A.2)-(A.3) it follows that there exists $1 \leq l \leq N - 1$ so that $\{k_1, k_2, k_3, k_4\} = \{l, -N+l, m, n\}$ for some $m, n \in \mathbb{Z}$. Then $\sin \frac{l\pi}{N} + \sin \frac{(-N+l)\pi}{N} = 0$, and hence by (A.1), $\sin \frac{m\pi}{N} + \sin \frac{n\pi}{N} = 0$. W.l.o.g. assume that $1 \leq m \leq N - 1$. Then either $n = -m$ or $n = -N + m$. If $n = -m$, then $(k_1, k_2, k_3, k_4) \in K_4^{res}$ by Lemma A.3 (i). If $n = -N + m$, then one has

$$\sum_{i=1}^{4} k_i = 2l - N + 2m - N = 2(l+m) - 2N.$$

Note that $2(l+m) - 2N$ cannot be an even multiple of N, as otherwise $l+m \equiv 0 \mod N$, violating (A.2). If, in addition, N is odd, then $2(l+m) - 2N$ cannot be an odd multiple of N. Hence in the case where N is odd we conclude that $\sum_{i=1}^{4} k_i \not\equiv 0 \mod N$, contradicting the assumption $(k_1, k_2, k_3, k_4) \in K_4$.

If N is even, it is however possible that $2(l+m) - 2N$ equals $\pm N$: If $2(l+m) - 2N = N$, i.e. $l + m = \frac{3}{2}N$, it follows that $\frac{N}{2} < l, m \leq N-1$, and

$(k_1, k_2, k_3, k_4) \in K_{res}^-$ as $\{k_1, k_2, k_3, k_4\} = \{-l', -l' + N, \frac{N}{2} + l', -\frac{N}{2} + l'\}$ with $l' = l - \frac{N}{2}$. If $2(l + m) - 2N = -N$, i.e. $l + m = \frac{N}{2}$, it follows similarly that $(k_1, k_2, k_3, k_4) \in K_{res}^+$ as $\{k_1, k_2, k_3, k_4\} = \{l, l - N, \frac{N}{2} - l, -\frac{N}{2} - l\}$. So in both cases, we conclude that $(k_1, k_2, k_3, k_4) \in K_4^{res}$. □

In view of Lemma A.3 and Lemma A.4, in order to prove Lemma A.1 it remains to show the following

Lemma A.5. *Assume that $(k_1, k_2, k_3, k_4) \in K_4$ satisfies (A.2). If for any $1 \le i, j \le 4$*

$$\sin \frac{k_i \pi}{N} + \sin \frac{k_j \pi}{N} \ne 0. \tag{A.4}$$

(and thus $(k_1, k_2, k_3, k_4) \notin K_4^N \cup K_4^{res}$), then

$$\sum_{i=1}^{4} \sin \frac{k_i \pi}{N} \ne 0.$$

To prove Lemma A.5 let us first use Euler's formula for the sine function to rewrite (A.1) as

$$\sum_{1 \le |j| \le 4} \zeta_j = 0, \tag{A.5}$$

where $\zeta_{\pm j} = \pm e^{\pm i k_j \pi / N}$ are $2N$'th roots of unity. Note that for any quadruple $(k_1, k_2, k_3, k_4) \in K_4 \setminus K_4^N$ satisfying (A.4) one has

$$\zeta_j + \zeta_{j'} \ne 0 \quad \forall\, 1 \le |j| \le |j'| \le 4.$$

Indeed for any $1 \le |j| \le |j'| \le 4$ one has $\operatorname{Im} \zeta_j + \operatorname{Im} \zeta_{j'} = \sin \frac{k_{|j|}\pi}{N} + \sin \frac{k_{|j'|}\pi}{N}$ which does not vanish by the assumption (A.4).

Let us first discuss equation (A.5) and its solutions in general, i.e. we consider the equation

$$\zeta_1 + \ldots + \zeta_8 = 0 \tag{A.6}$$

and study its solutions, $(\zeta_l)_{1 \le l \le 8}$, on the unit circle $S^1 := \{z \in \mathbb{C} \mid |z| = 1\}$.

We need an auxiliary result which we discuss first. Let $n \ge 2$ be arbitrary and assume that the sequence $(\zeta_i)_{1 \le i \le n} \subseteq S^1$ satisfies the equation

$$\sum_{i=1}^{n} \zeta_i = 0, \tag{A.7}$$

but no vanishing subsums (i.e. $\sum_{l \in J} \zeta_l \ne 0$ for any $\emptyset \subsetneq J \subsetneq \{1, \ldots, n\}$).

Let $M \in \mathbb{N}$ be the smallest positive integer with the property that $(\zeta_i/\zeta_j)^M = 1$ for all $1 \le i, j \le n$. Then there exists $\xi \in S^1$ so that $\zeta_i^M = \xi^M$ for any $1 \le i \le n$. W.l.o.g. we can assume that $\xi = 1$. Furthermore, let p^k be a prime power dividing M so that M/p^k and p are relatively prime, and define

$$M' =: M/p, \quad \eta := e^{2\pi i/p^k}. \tag{A.8}$$

Then for any $1 \leq l \leq n$ there exists a unique integer $\mu(l)$ with $0 \leq \mu(l) \leq p-1$ such that $\zeta_l = \tilde{\zeta}_l \cdot \eta^{\mu(l)}$, where $\tilde{\zeta}_l$ is an element of the field $K := \mathbb{Q}(e^{2\pi i/M'})$; this establishes a map
$$\mu : \{1,\ldots,n\} \to \{0,\ldots,p-1\}. \tag{A.9}$$
(As $\zeta_l^M = 1$ there exists $0 \leq r_l \leq M-1$ with $\zeta_l = e^{\frac{2\pi i}{M}r_l}$. If $r_l \equiv 0 \bmod p$ choose $\mu(l) = 0$. If $r_l \not\equiv 0 \bmod p$ choose $1 \leq \mu(l) \leq p-1$ so that $r_l \equiv \frac{M}{p^k}\mu(l) \bmod p$.) Hence (A.7) can be written as

$$0 = \sum_{l=1}^{n} \zeta_l = \sum_{s=0}^{p-1}\left(\sum_{l\in\mu^{-1}(s)} \zeta_l\right) = \sum_{s=0}^{p-1}\left(\sum_{l\in\mu^{-1}(s)} \tilde{\zeta}_l\right)\eta^s. \tag{A.10}$$

We need the following algebraic fact (see e.g. [95], p. 60-61):

Proposition A.6. *The minimal polynomial of $\eta = e^{2\pi i/p^k}$ over the field $K = \mathbb{Q}(e^{2\pi i/M'})$ is given by $X^p - \eta^p$ if $k \geq 2$ and $X^{p-1} + X^{p-2} + \ldots + X + 1$ if $k = 1$.*

We now claim that M is square-free, or equivalently, that for any prime power p^k dividing M,
$$k = 1. \tag{A.11}$$
Indeed, equation (A.10) shows that the minimal polynomial of ζ has degree at most $p-1$, which by Proposition A.6 is only satisfied in the case $k=1$.

Further we claim that there exists $\sigma \in \mathbb{C} \setminus \{0\}$ so that
$$\sum_{l\in\mu^{-1}(s)} \tilde{\zeta}_l = \sigma \quad \forall 0 \leq s \leq p-1. \tag{A.12}$$

The existence of such a σ follows from Proposition A.6: As $k = 1$ by (A.11), the minimal polynomial of η over K is given by $X^{p-1} + X^{p-2} + \ldots + X + 1$. Since this is a polynomial of degree $p-1$, the polynomial on the right hand side of (A.10) must be a scalar multiple of the minimal polynomial. Hence all coefficients $\sum_{l\in\mu^{-1}(s)} \tilde{\zeta}_l$ have the same value $\sigma \in \mathbb{C}$. As $\sum_{l\in\mu^{-1}(s)} \zeta_l = \sigma\eta^s$ the additional property $\sigma \neq 0$ follows from the assumption that there are no vanishing subsums. Hence we can assume w.l.o.g. that $\sigma = 1$.

Next we claim that
$$p \leq n. \tag{A.13}$$
In other words, possible prime factors of M are bounded by the number of summands in (A.7). To prove (A.13), note that it follows from (A.12) that for any $0 \leq s \leq p-1$ there exists $1 \leq l \leq n$ such that $\mu(l) = s$, i.e. the map μ, defined by (A.9), is onto. This establishes (A.13).

The map μ induces the *partition* $(\sharp\mu^{-1}(s))_{0\leq s\leq p-1}$ of the positive integer n into p summands,
$$n = \sum_{s=0}^{p-1} \sharp\mu^{-1}(s). \tag{A.14}$$

Lemma A.7. *For any solution $\{\zeta_1, \ldots, \zeta_8\}$ of (A.6) contained in S^1 without vanishing subsums there exists $\xi \in S^1$ such that either*

$$\{\zeta_1, \ldots, \zeta_8\} = \{-\xi\alpha, -\xi\alpha^2\} \cup \{\xi\gamma^j \mid 1 \leq j \leq 6\} \tag{A.15}$$

or

$$\{\zeta_1, \ldots, \zeta_8\} = \{-\xi\alpha, -\xi\alpha^l \cdot \beta^i, -\xi\alpha^l \cdot \beta^j \mid 1 \leq l \leq 2\} \cup \{\xi\beta^k, \xi\beta^m\}, \tag{A.16}$$

where the quadruple (i, j, k, m) is a permutation of $(1, 2, 3, 4)$ and

$$\alpha := e^{\frac{2\pi i}{3}}, \quad \beta := e^{\frac{2\pi i}{5}}, \quad \gamma := e^{\frac{2\pi i}{7}}.$$

Proof. By a straightforward computation one verifies that the sets of the form (A.15) and (A.16) satisfy (A.6). It remains to prove that these are the only solutions of (A.6) of this type.

We classify the solutions of (A.6) according to the possible values of p, which we now assume to be the largest prime dividing M. Since $n = 8$, by (A.13), the possible values of p are 2, 3, 5, and 7. If $p = 2$, then, by (A.11), $M = 2$ and therefore there exists $\xi \in S^1$ so that $\zeta_j = \pm \xi$ for any $1 \leq j \leq n$. In this case there exists a solution of (A.7) without vanishing subsums only if $n = 2$. (In this case, they are given by $\{\zeta_1, \zeta_2\} = \xi\{1, -1\}$ with $\xi \in S^1$.) If $p = 3$, then $M = 3$ or $M = 3 \cdot 2$, and there exists a solution of (A.7) without vanishing subsums only if $n = 3$. (In this case, they are given by $\{\zeta_1, \zeta_2, \zeta_3\} = \xi\{1, \alpha, \alpha^2\}$ with $\xi \in S^1$.) If $p = 5$, then the quantity η defined in (A.8) is given by $\eta = \beta = e^{2\pi i/5}$. Up to permutations, there are the following three partitions of 8 into 5 summands, $(4, 1, 1, 1, 1)$, $(3, 2, 1, 1, 1)$, and $(2, 2, 2, 1, 1)$. In a straightforward way one shows that the partitions $(4, 1, 1, 1, 1)$ and $(3, 2, 1, 1, 1)$ and their permutations give rise to solutions of the equation (A.6) with vanishing subsums. E.g. the solutions corresponding to $(4, 1, 1, 1, 1)$ are given by $\xi \cdot (-\beta, -\beta^2, -\beta^3, -\beta^4, \beta, \beta^2, \beta^3, \beta^4)$ with $\xi \in S^1$, whereas the solutions corresponding to $(3, 2, 1, 1, 1)$ are $\xi \cdot (-i, 1, i, -\alpha\beta, -\alpha^2\beta, \beta^2, \beta^3, \beta^4)$ with $\xi \in S^1$. On the other hand the partition $(2, 2, 2, 1, 1)$ leads to the solutions

$$(\zeta_1, \ldots, \zeta_8) = \xi(-\alpha, -\alpha^2, -\alpha\beta, -\alpha^2\beta, -\alpha\beta^2, -\alpha^2\beta^2, \beta^3, \beta^4)$$

with $\xi \in S^1$. They are the solutions (A.16) with $(i, j, k, m) = (1, 2, 3, 4)$. Permutations of the partition $(2, 2, 2, 1, 1)$ again lead to solutions of the type (A.16), but with (i, j, k, m) given by a permutation of $(1, 2, 3, 4)$.

If $p = 7$, then $\eta = \gamma$ in (A.8). Then, up to permutations, $(2, 1, 1, 1, 1, 1, 1)$ is the only possible partition of 8 into 7 summands. The partition $(2, 1, 1, 1, 1, 1, 1)$ leads to the solutions

$$(\zeta_1, \ldots, \zeta_8) = \xi(-\alpha, -\alpha^2, \gamma, \ldots, \gamma^6)$$

with $\xi \in S^1$, where we used that $1 = -\alpha - \alpha^2$. They are of type (A.15). Any permutation of $(2, 1, 1, 1, 1, 1, 1)$ leads to the same kind of solutions. □

Lemma A.8. *For any solution $\{\zeta_1, \ldots, \zeta_8\}$ of (A.6) contained in S^1 without vanishing subsums of length 2 but having a vanishing subsum of length 3, 4, or 5, there exist $\xi, \xi' \in S^1$ such that*

$$\{\zeta_1, \ldots, \zeta_8\} = \{\xi \alpha^l | 0 \leq l \leq 2\} \cup \{\xi' \beta^m | 0 \leq m \leq 4\}, \quad (A.17)$$

where again $\alpha = e^{2\pi i/3}$ and $\beta = e^{2\pi i/5}$.

Proof. Again, one verifies by a direct computation that the solutions (A.17) of (A.6) have the desired properties. It remains to prove that they are the only ones. First note that under the hypotheses of the lemma, vanishing subsums of length 4 cannot occur, since the latter ones would imply the existence of vanishing subsums of length 2, which by assumption is excluded. Hence, in order to find solutions of (A.7) for $n = 8$ with the desired properties, we have to find all solutions of (A.7) without vanishing subsums for $n = 3$ and $n = 5$. Note that by (A.13), $p = n$ for $n = 3$ or $n = 5$. By the considerations in the proof of Lemma A.7, the former ones are given by $(\zeta_1, \zeta_2, \zeta_3) = \xi(1, \alpha, \alpha^2)$ and the latter ones by $(\zeta_1, \ldots, \zeta_5) = \xi'(1, \beta, \beta^2, \beta^3, \beta^4)$ with $\xi, \xi' \in S^1$. This proves the lemma. □

We are now ready to prove Lemma A.5.

Proof of Lemma A.5. We first select from (A.15), (A.16) and (A.17) all solutions $(\zeta_1, \ldots, \zeta_8)$ of (A.6) which are of the form (A.1) (after multiplication by $2i$). This amounts to selecting the solutions $(\zeta_1, \ldots, \zeta_8)$ of (A.6) having the property that $\{\zeta_1, \ldots, \zeta_8\}$ is invariant under the map $\zeta \mapsto -\zeta^{-1}$. This requires to choose ξ and ξ' in (A.15), (A.16), and (A.17) appropriately. Let us explain this procedure in detail for the solutions of type (A.15).

First we rewrite the solution (A.15),

$$(\zeta_1, \ldots, \zeta_8) = \xi \cdot (-\alpha, -\alpha^2, \gamma, \gamma^2, \gamma^3, \gamma^4, \gamma^5, \gamma^6) = e^{\frac{2\pi i x}{42}} \left(e^{\frac{2\pi i t_k}{42}} \right)_{1 \leq k \leq 8},$$

where $\xi = e^{2\pi i x/42}$ with $x \in \mathbb{R}/42\mathbb{Z}$ and

$$(t_1, \ldots, t_8) = (6, 7, 12, 18, 24, 30, 35, 36). \quad (A.18)$$

The required invariance of the set of the ζ_k's under the map $\zeta \mapsto -\zeta^{-1}$ is equivalent to the invariance of the set of the $(t_k + x)$'s under the map $t \mapsto 21 - t$ (mod 42). Since the set (A.18) of the t_k's is invariant under the map $t \mapsto -t$ (mod 42), $\{t_k + x | 1 \leq k \leq 8\}$ is invariant under $t \mapsto 21 - t$ (mod 42), if we choose $x := \frac{21}{2}$ or $\xi = i$. Then the equation $\sum_{i=1}^{8} \zeta_i = 0$ reads

$$e^{\frac{11\pi i}{14}} + e^{\frac{5\pi i}{6}} + e^{\frac{15\pi i}{14}} + e^{\frac{19\pi i}{14}} + e^{\frac{23\pi i}{14}} + e^{\frac{27\pi i}{14}} + e^{\frac{\pi i}{6}} + e^{\frac{3\pi i}{14}} = 0,$$

or $\sin \frac{\pi}{6} + \sin \frac{3\pi}{14} + \sin \frac{15\pi}{14} + \sin \frac{19\pi}{14} = 0$. Choosing all arguments in $(0, \pi)$, the latter identity reads

$$\sin \frac{\pi}{6} + \sin \frac{3\pi}{14} - \sin \frac{\pi}{14} - \sin \frac{5\pi}{14} = 0. \quad (A.19)$$

For the solutions of type (A.16), one gets

$$\sin\frac{\pi}{6} + \sin\frac{13\pi}{30} - \sin\frac{7\pi}{30} - \sin\frac{3\pi}{10} = 0 \qquad (A.20)$$

and

$$\sin\frac{\pi}{6} + \sin\frac{\pi}{30} - \sin\frac{11\pi}{30} + \sin\frac{\pi}{10} = 0. \qquad (A.21)$$

Let us briefly explain how (A.20)-(A.21) can be obtained. Note that from the 24 permutations of $(1,2,3,4)$ in (A.16), there are only six which lead to different sets of the ζ_i's, since interchanging i and j or k and m leaves the set on the right hand side of (A.16) invariant. In the resulting six different cases, we again write $\{\zeta_1,\ldots,\zeta_8\} = \xi \cdot \{e^{2\pi i \cdot \frac{t_1}{30}},\ldots,e^{2\pi i \cdot \frac{t_8}{30}}\}$ with t_k in $\mathbb{R}/30\mathbb{Z}$. Then, up to translations, there are only two different types of solutions emerging from these six cases. With the appropriate choices of ξ, one gets the solutions (A.20) and (A.21).

Finally, for the solutions of type (A.17), one gets

$$\sin\frac{\pi}{2} - \sin\frac{\pi}{6} + \sin\frac{\pi}{10} - \sin\frac{3\pi}{10} = 0. \qquad (A.22)$$

The procedure to obtain (A.22) is basically the same as in the preceding cases. We write (A.17) as $\{\zeta_1,\ldots,\zeta_8\} = \xi \cdot \{\alpha^l, \lambda\cdot\beta^m | 0 \leq l \leq 2,\ 0 \leq m \leq 4\}$ and first choose $\lambda \in S^1$ so that the set $\{\alpha^l, \lambda\cdot\beta^m | 0 \leq l \leq 2,\ 0 \leq m \leq 4\}$ is symmetric with respect to some axis through the origin, and then choose ξ so that this axis is the imaginary axis.

To finish the proof of Lemma A.5, it remains to show that all the solutions (k_1,k_2,k_3,k_4) of (A.1) obtained in (A.19)-(A.22) and the additional ones obtained by replacing $0 < x < \pi$ in $\sin x$ by $\pi - x$ satisfy $\sum_{i=1}^{4} k_i \not\equiv 0 \bmod N$ and hence are not contained in K_4.

For the solutions obtained in (A.19)-(A.22), N is even. Hence if N is odd, there is no quadruple $(k_1,k_2,k_3,k_4) \in K_4$ such that (A.1) and (A.4) are satisfied. This finishes the proof of Lemma A.5 in this case.

For the rest of the proof, we assume that N is even. If $N = 42r$ for some $r \in \mathbb{N}$, (A.19) becomes

$$\sin\frac{7r\pi}{42r} + \sin\frac{9r\pi}{42r} + \sin\frac{(-3r)\pi}{42r} + \sin\frac{(-15r)\pi}{42r} = 0,$$

and we have $7r + 9r - 3r - 15r = -2r \not\equiv 0 \bmod 42r$. Hence the corresponding quadruple (k_1,k_2,k_3,k_4) is not contained in K_4. For the quadruples obtained by replacing $0 < x < \pi$ in $\sin x$ by $\pi - x$ in some of the summands in (A.19), the condition $\sum_{i=1}^{4} k_i \not\equiv 0 \bmod 42r$ amounts to

$$\pm 7 \pm 9 \pm 3 \pm 15 \not\equiv 0 \bmod 42 \qquad (A.23)$$

for any combination of plus and minus signs. The relations (A.23) are easily verified. Similarly, one verifies that the quadruples (k_1,k_2,k_3,k_4) satisfying (A.20), (A.21), or (A.22) are not in K_4 by showing that

$$\pm 5 \pm 13 \pm 7 \pm 9 \not\equiv 0, \quad \pm 5 \pm 1 \pm 11 \pm 3 \not\equiv 0, \quad \pm 15 \pm 5 \pm 3 \pm 9 \not\equiv 0 \bmod 30, \qquad (A.24)$$

again for any combination of plus and minus signs. Hence we have shown that none of the solutions (k_1, k_2, k_3, k_4) of (A.1) is an element of K_4. This completes the proof of Lemma A.5. □

Proof of Lemma A.1. The claimed statement follows from Lemmas A.2, A.3, A.4, and A.5. □

Appendix B

Lemma on Symmetric Polynomials

Here we prove Lemma 4.1.5 on the values of the N elementary symmetric polynomials in $N-1$ variables for $t_k = \sin^2 \frac{k\pi}{N}$, $1 \leq k \leq N-1$. Recall from (4.19)-(4.20) that these polynomials are defined by

$$\Pi_0 := 1, \tag{B.1}$$

$$\Pi_n(t_1, \ldots, t_{N-1}) := \sum_{1 \leq i_1 < \ldots < i_n \leq N-1} t_{i_1} \cdot \ldots \cdot t_{i_n} \quad (1 \leq n \leq N-1). \tag{B.2}$$

We restate Lemma 4.1.5 as follows:

Lemma B.1. *Let $N \geq 2$ be an arbitrary integer (not necessarily odd). Evaluated at $t_k = \sin^2 \frac{k\pi}{N}$ $(1 \leq k \leq N-1)$, the N elementary symmetric polynomials $(\Pi_n)_{0 \leq n \leq N-1}$ given by (B.1) and (B.2) take the values*

$$\Pi_n\left(\sin^2 \frac{\pi}{N}, \ldots, \sin^2 \frac{(N-1)\pi}{N}\right) = 4^{-n} \frac{N}{N-n} \binom{2N-n-1}{n}. \tag{B.3}$$

Before proving Lemma B.1, we turn to the following (seemingly unmotivated) combinatorial fact:

Lemma B.2. *For any $N, m \in \mathbb{N}$,*

$$\sum_{k=0}^{m} \frac{(-1)^k}{N+k} \binom{2k}{k} \binom{2N+m+k-1}{m-k} = \frac{1}{N+m} \binom{2N+m-1}{m}. \tag{B.4}$$

Proof. To prove (B.4), we proceed inductively in m for any fixed N. First note that the identity holds for $m = 0$ and $m = 1$, since in these cases both sides of (B.4) are equal to $\frac{1}{N}$ and $\frac{2N}{N+1}$, respectively. For the induction step, we claim that both sides of (B.4) satisfy the recurrence relation

$$(N+m)(2N+m)f_m - (m+1)(N+m+1)f_{m+1} = 0 \tag{B.5}$$

for any $m \in \mathbb{N}$. That the right hand side of (B.4) satisfies (B.5) can be checked by a direct computation. To see that the same holds for the left hand side of (B.4), first note that the summand

$$F_{k,m} := \frac{(-1)^k}{N+k}\binom{2k}{k}\binom{2N+m+k-1}{m-k}$$

of the left hand side of (B.4) satisfies the recurrence relation

$$(N+m)(2N+m)F_{k,m} - (m+1)(N+m+1)F_{k,m+1} = \Delta_k(F_{k,m}R_{k,m}), \quad (B.6)$$

where for $k \leq m$

$$R_{k,m} := \frac{k(N+k)(2N+2k-1)}{m-k+1},$$

and where Δ_k is the standard discrete difference operator of order one in k, i.e. $\Delta_k(F_{k,m}R_{k,m}) = F_{k+1,m}R_{k+1,m} - F_{k,m}R_{k,m}$. The proof that $F_{k,m}$ satisfies (B.6) is a straightforward (lengthy) computation.

We now denote the left hand side of (B.4) by f_m, i.e. for any $m \in \mathbb{N}$

$$f_m := \sum_{k=0}^{m} F_{k,m} = \sum_{k=0}^{m} \frac{(-1)^k}{N+k}\binom{2k}{k}\binom{2N+m+k-1}{m-k}.$$

Adding up (B.6) for k from 0 to $m-1$, we get

$$(N-m)(2N+m)(f_m - F_{m,m}) - (m+1)(N+m+1)(f_{m+1} - (F_{m,m+1} + F_{m+1,m+1}))$$
$$= F_{m,m}R_{m,m} - F_{0,m}R_{0,m} \quad (B.7)$$

Further, by a direct computation we obtain

$$F_{m,m}R_{m,m} - F_{0,m}R_{0,m} = (-1)^m m(2N+2m-1)\binom{2m}{m}, \quad (B.8)$$

Another (lengthy) computation shows that

$$(N+m)(2N+m)F_{m,m} - (m+1)(N+m+1)(F_{m,m+1} + F_{m+1,m+1})$$
$$= -(-1)^m m(2N+2m-1)\binom{2m}{m}.$$

Adding this to (B.7) and combining the result with (B.8) yields the recurrence relation (B.5) and hence completes the proof of the combinatorial fact (B.4). □

Remark: The recurrence relations (B.5) and (B.6) were found using the Mathematica program zb-alg.m written by P. Paule, M. Schorn, & A. Riese, which is an implementation of D. Zeilberger's "creative telescoping" algorithm, described e.g. in Chapter 6 of [72]. We however emphasize that the *proof* of both recurrence relations is purely analytical, the program just mentioned was "only" used to *find* the recurrence relations.

Note that by setting $N+m =: N'$ (after which the $'$ is again omitted), (B.4) is equivalent to the fact that for any $N \geq 2$ and $1 \leq m \leq N-1$,

$$\sum_{k=0}^{m} \frac{(-1)^k}{N-(m-k)} \binom{2k}{k} \binom{2N-(m-k)-1}{m-k} = \frac{1}{N}\binom{2N-m-1}{m}. \quad \text{(B.9)}$$

Omitting the summand $k=0$ in (B.9), using the formula

$$\frac{1}{r}\binom{r}{l} = \frac{1}{r-1}\binom{r-1}{l} \qquad (r,l \in \mathbb{N},\ l < r) \quad \text{(B.10)}$$

for $r := 2N-(m-k)$ and $l := m-k$ and dividing by 2, we obtain

$$\sum_{k=1}^{m} \frac{(-1)^k}{2N-(m-k)} \binom{2k}{k} \binom{2N-(m-k)}{m-k} = -\frac{m}{2N(N-m)}\binom{2N-m-1}{m}. \quad \text{(B.11)}$$

Proof of Lemma B.1. For fixed $N \geq 2$, we proceed by induction on n. For $n = 0, 1$, (B.3) is easily verified. For the induction step, we use the formulas of Newton-Girard (see e.g. [89], p. 278-279) expressing for arbitrary $t_1, \ldots, t_{N-1}$ the polynomials $\Pi_n(t_1, \ldots, t_{N-1})$ inductively through the Newton sums

$$S_n(t_1, \ldots, t_{N-1}) := \sum_{k=1}^{N-1} t_k^n. \quad \text{(B.12)}$$

Omitting the arguments, these Newton-Girard formulas are given by

$$m \Pi_m + \sum_{k=1}^{m} (-1)^k S_k \Pi_{m-k} = 0, \quad 1 \leq m \leq N-1. \quad \text{(B.13)}$$

To apply (B.13), we first cite from ([79], p. 640) the value of the Newton sums (B.12) for $t_k = \sin^2 \frac{k\pi}{N}$, namely

$$S_n\left(\sin^2 \frac{\pi}{N}, \ldots \sin^2 \frac{(N-1)\pi}{N}\right) \equiv \sum_{k=1}^{N-1} \sin^{2n}\frac{k\pi}{N} = \frac{N}{4^n}\binom{2n}{n}. \quad \text{(B.14)}$$

Assume now that (B.3) is verified for all $0 \leq m \leq n$ and that $n+1 \leq N-1$. We again use the identity (B.10) for $r := 2N-n$ and $l := n$ to observe that the right hand side of (B.3) is equal to $4^{-n}\frac{2N}{2N-n}\binom{2N-n}{n}$. Substituting (B.3) and (B.14) into the Newton-Girard formula (B.13) with $m := n+1$, i.e.

$$\Pi_{n+1} = -\frac{1}{n+1}\sum_{k=1}^{n+1}(-1)^k S_k \Pi_{n+1-k},$$

we get for $\Pi_{n+1} = \Pi_{n+1}\left(\sin^2 \frac{\pi}{N}, \ldots, \sin^2 \frac{(N-1)\pi}{N}\right)$

$$\Pi_{n+1} = -\frac{1}{n+1}\sum_{k=1}^{n+1}(-1)^k \frac{N \cdot 4^{-(k+(n+1-k))} \cdot 2N}{2N-(n+1-k)}\binom{2k}{k}\binom{2N-(n+1-k)}{n+1-k}$$

$$= -\frac{2N^2 \cdot 4^{-(n+1)}}{n+1}\sum_{k=1}^{n+1}\frac{(-1)^k}{2N-(n+1-k)}\binom{2k}{k}\binom{2N-(n+1-k)}{n+1-k}$$

By (B.11), it follows that

$$\Pi_{n+1} = 4^{-(n+1)} \frac{N}{N-(n+1)} \binom{2N-(n+1)-1}{n+1}.$$

This completes the induction step and therefore the proof of Lemma B.1. □

Bibliography

[1] V. I. ARNOL'D, V. V. KOZLOV, & A. I. NEISHTADT, Mathematical Aspects of Classical and Celestial Mechanics. In: V. I ARNOL'D (ED.), *Dynamical Systems II*, 2nd ed., Encyclopedia of Mathematical Sciences vol. 3, Springer, Berlin, 1993, 1-291.

[2] V. I. ARNOL'D, *Mathematical Methods of Classical Mechanics*. 2nd ed., Graduate Texts in Mathematics **60**, Springer New York, 1989.

[3] V. I. ARNOL'D, Proof of a theorem by A. N. Kolmogorov on the invariance of quasi-periodic motions under small perturbations of the Hamiltonian. *Russian Math. Surveys* **18** (1963), 9-36.

[4] V. I. ARNOL'D & A. AVEZ, *Ergodic Problems of Classical Mechanics*. W. A. Benjamin, New York, 1968.

[5] D. BAMBUSI, T. KAPPELER, & T. PAUL, De Toda à KdV. *C. R. Acad. Sci. Paris* Ser. I **340** (2005).

[6] D. BAMBUSI & A. PONNO, Korteweg-de Vries equation and energy sharing in Fermi-Pasta-Ulam. *Chaos* **15** (2005), 015107.

[7] D. BAMBUSI & A. PONNO, On metastability in FPU. *Comm. Math. Phys.* **264** (2006), 539-561.

[8] G. P. BERMAN & F. M. IZRAILEV, The Fermi-Pasta-Ulam problem: Fifty years of progress. *Chaos* **15** (2005), 015104.

[9] G. D. BIRKHOFF, *Dynamical Systems*. AMS Colloquium Publications, Vol. IX, 1927 [revised edition: 1966].

[10] H. W. BROER, KAM theory: the legacy of Kolmogorov's 1954 paper. *Bull. AMS (New Series)* **41**(4) (2004), 507-521.

[11] H. W. BROER, G. B. HUITEMA, & M. B. SEVRYUK, *Quasi-periodic Motions in Families of Dynamical Systems: Order Amidst Chaos*. Lecture Notes in Mathematics, Vol. 1645, Springer, Berlin, 1996.

[12] T. BOUNTIS, H. SEGUR, & F. VIVALDI, Integrable Hamiltonian systems and the Painlévé property. *Phys. Rev. A* **25**(3) (1982), 1257-1264.

[13] D. K. CAMPBELL, P. ROSENAU, & G. M. ZASLAVSKY, Introduction: The Fermi-Pasta-Ulam problem - The first fifty years. *Chaos* **15** (2005), 015101.

[14] A. CARATI, L. GALGANI, & A. GIORGILLI, The Fermi-Pasta-Ulam problem as a challenge for the foundations of physics. *Chaos* **15** (2005), 015105.

[15] J. H. CONWAY & A. J. JONES, Trigonometric diophantine equations (On vanishing sums of roots of unity). *Acta Arithmetica* **XXX** (1976), 229-240.

[16] R. H. CUSHMAN & L. M. BATES, *Global Aspects of Classical Integrable Systems*. Birkhäuser, Basel, 1997.

[17] T. DAUXOIS, Fermi, Pasta, Ulam and a mysterious lady. *Physics Today* **61**(1), 55-57, 2008.

[18] T. DAUXOIS, M. PEYRARD & S. RUFFO, The Fermi-Pasta-Ulam "numerical experiment": history and pedagogical perspectives. *Eur. J. Phys.* **26** (2005), 3-11.

[19] F. FASSÒ, M. GUZZO & G. BENETTIN, Nekhoroshev-stability of elliptic equilibria of Hamiltonian systems. *Comm. Math. Phys.* **197** (1998), 347-360.

[20] F. FASSÒ, M. GUZZO & G. BENETTIN, Nekhoroshev-stability of L4 and L5 in the spatial restricted three–body problem. *Reg. Chaot. Dyn.* **3** (3) (1998), 56-71.

[21] H. FLASCHKA, The Toda lattice. I. Existence of integrals. *Phys. Rev.*, Sect. B **9** (1974), 1924-1925.

[22] E. FERMI, Beweis, dass ein mechanisches Normalsystem im allgemeinen quasi-ergodisch ist. *Phys. Zeitschrift* **24** (1923), 261-265.

[23] E. FERMI, J. PASTA & S. ULAM, Studies of non linear problems. *Los Alamos Rpt.* **LA-1940** (1955). In: *Collected Papers of Enrico Fermi*. University of Chicago Press, Chicago, 1965, Volume II, 978-988. Theory, Methods and Applications, 2nd ed., Marcel Dekker, New York, 2000. Also contained in: S. M. ULAM, *Analogies between Analogies. The mathematical reports of S. M. Ulam and his Los Alamos collaborators*. Edited by A. R. Bednarek & F. Ulam. University of California Press, Berkeley, 1990.

[24] J. FORD, Equipartition of energy for nonlinear systems. *J. Math. Phys.* **2**(3) (1961), 387-393.

[25] J. FORD & J. WATERS, Computer studies of energy sharing and ergodicity for nonlinear oscillator systems. *J. Math. Phys.* **4**(10) (1963), 1293-1306.

[26] J. FORD, The Fermi-Pasta-Ulma problem: paradox turns discovery. *Physics Reports* **213**(5) (1992), 271-310.

[27] G. FRIESECKE & R. L. PEGO, Solitary waves on FPU lattices: I. Qualitative properties, renormalization and continuum limit. *Nonlinearity* **12** (1999), 1601-1627.

[28] G. FRIESECKE & R. L. PEGO, Solitary waves on FPU lattices: II. Linear implies nonlinear stability. *Nonlinearity* **15** (2002), 1343-1359.

[29] G. FRIESECKE & R. L. PEGO, Solitary waves on Fermi-Pasta-Ulam lattices: III. Howland-type Floquet theory. *Nonlinearity* **17** (2004), 207-227.

[30] G. FRIESECKE & R. L. PEGO, Solitary waves on Fermi-Pasta-Ulam lattices: IV. Proof of stability at low energy. *Nonlinearity* **17** (2004), 229-251.

[31] F. FUCITO, F. MARCHESONI, E. MARINARI, G. PARISI, L. PELITI, S. RUFFO & A. VULPIANI, Approach to equilibrium in a chain of nonlinear oscillators. *J. de Physique* **43** (1982), 707-713.

[32] P. GALISON, *How Experiments End.* University of Chicago Press, Chicago, 1987.

[33] C. S. GARDNER, J. M. GREENE, M. D. KRUSKAL & R. M. MIURA, Method for solving the Korteweg-de Vries equation. *Phys. Rev. Lett.* **19** (1967), 1095-1097.

[34] C. S. GARDNER, J. M. GREENE, M. D. KRUSKAL & R. M. MIURA, Korteweg-de Vries equation and generalizations. II. Existence of conservation laws and constants of motion. *J. Math. Phys.* **9** (1968), 1204-1209.

[35] B. GRAMMATICOS, A. RAMANI & V. PAPAGIORGIOU, Do integrable mappings have the Painléve property? *Phys. Rev. Lett.* **67**(14) (1991), 1825-1828.

[36] F. G. GUSTAVSON, On constructing formal integrals of a Hamiltonian system near an equilibrium point. *Astronom. J.* **71**(8) (1966), 670-686.

[37] M. HÉNON, Integrals of the Toda lattice. *Phys. Rev.*, Sect. B **9** (1974), 1921.

[38] M. HÉNON & C. HEILES, The applicability of the third integral of motion: Some numerical experiments. *Astronom. J.* **69**(1) (1964), 73-79.

[39] A. HENRICI, *Normal Form for Fermi-Pasta-Ulam Chains.* Dissertation, Universität Zürich, 2008.

[40] A. HENRICI & T. KAPPELER, Results on normal forms for FPU chains. *Comm. Math. Phys.* **278** (2008), 145-177.

[41] A. HENRICI & T. KAPPELER, Global action-angle variables for the periodic Toda lattice. *Int. Math. Res. Not.* 2008; Vol. 2008: article ID rnn031, 52 pages, DOI: 10.1093/imrn/rnn031.

[42] A. HENRICI & T. KAPPELER, Global Birkhoff coordinates for the periodic Toda lattice. *Nonlinearity* **21** (2008), 2731-2758.

[43] A. HENRICI & T. KAPPELER, Birkhoff normal form for the periodic Toda lattice. In: Integrable Systems and Random Matrices, *Contemp. Math.* **458** (2008), 11-29.

[44] A. HENRICI & T. KAPPELER, Nekhoroshev theorem for the periodic Toda lattice. *Chaos* **19** (2009), 033120.

[45] A. HENRICI & T. KAPPELER, Resonant normal form for even periodic FPU chains. *J. Eur. Math. Soc.* **11**(5) (2009), 1025-1056.

[46] YU. S. IL'YASCHENKO, A steepness test for analytic functions. *Uspekhi Mat. Nauk* **41** (1986), 193-194 [Russian]. English translation in: *Russian math. Surveys* **41** (1986), 229-230.

[47] F. M. IZRAILEV & B. V. CHIRIKOV, Statistical properties of a nonlinear string. *Doklady Akademii Nauk SSSR* **166**(1) (1965) [Russian]. English translation in: *Soviet Phys. Dokl.* **11**(1) (1966), 30-32.

[48] E. A. JACKSON, Nonlinear coupled oscillators I: Perturbation theory; ergodic problem. *J. Math. Phys.* **4**(4) (1963), 551-558.

[49] E. A. JACKSON, Nonlinear coupled oscillators II: Comparison of theory with computer simulations. *J. Math. Phys.* **4**(5) (1963), 686-700.

[50] R. JOST, Winkel- und Wirkungsvariable für allgemeine mechanische Systeme. *Helvetica Physica Acta* **41** (1968), 965-968.

[51] T. KAPPELER & J. PÖSCHEL, *KdV & KAM*. Ergebnisse der Mathematik, 3. Folge, **45**, Springer, Berlin, 2003.

[52] A. N. KOLMOGOROV, On the conservation of conditionally periodic motions for a small change in Hamilton's function. *Dokl. Akad. Nauk SSSR* **98** (1954), 527-530 [Russian]. English translation in: *Lecture Notes in Physics* **45**, Springer, 1979.

[53] D. J. KORTEWEG, G. DE VRIES, On the change of form of long waves advancing in a rectangular canal, and on a new type of long stationary waves. *Phil. Mag. Ser. 5* **39** (1895), 422-443.

[54] T. KUHN, *The Structure of Scientific Revolutions*. 3rd ed., University of Chicago Press, Chicago, 1996.

[55] S. LANG, *Fundamentals of Differential Geometry*. Graduate Texts in Mathematics **191**, Springer, New York, 1999.

BIBLIOGRAPHY

[56] P. D. LAX, Integrals of nonlinear equations of evolution and solitary waves. *Comm. Pure Appl. Math.* **28** (1975), 141-188.

[57] R. DE LA LLAVE, A tutorial on KAM theory. *Proc. Symp Pure Math.* **69** (2001), 175-292.

[58] P. LOCHAK, Hamiltonian perturbation theory: periodic orbits, resonances and intermittency. *Nonlinearity* **6** (1993), 885-904.

[59] P. LOCHAK & A. NEISHTADT, Estimates of stability time for nearly integrable systems with a quasiconvex Hamiltonian. *Chaos* **2** (1992), DOI:10.1063/1.165891.

[60] T. LUCRETIUS CARUS, *On the Nature of the Universe* (Verse translation by James H. Mantinband). Ungar, New York, 1965.

[61] S. V. MANAKOV, Complete integrability and stochastization of discrete dynamical systems. *Zh. Exp. Teor. Fiz.* **67** (1974), 543-555 [Russian]. English translation: *Sov. Phys. JETP* **40** (1975), 269-274

[62] L. MARKUS & K. MEYER, Generic Hamiltonian dynamical systems are neither integrable nor ergodic. *Mem. Am. Math. Soc.* **144** (1974), 1-52.

[63] A. MORBIDELLI & A. GIORGILLI, Superexponential stability of KAM tori. *J. Stat. Phys.* **78** (1995), 1607-1617.

[64] J. MOSER, On invariant curves of area preserving mappings of an annulus. *Nachr. Akad. Wiss. Gött., Math. Phys. Kl.* (1962), 1-20.

[65] N. N. NEKHOROSHEV, Behaviour of Hamiltonian systems close to integrable. *Funct. Anal. Appl.* **5** (1971), 338-339.

[66] N. N. NEKHOROSHEV, An exponential estimate of the time of stability of nearly-integrable Hamiltonian systems I. *Uspekhi Mat. Nauk* **32** (1977), 5-66 [Russian]. English translation: *Russian Math. Surveys* **32** (1977), 1-65.

[67] N. N. NEKHOROSHEV, An exponential estimate of the time of stability of nearly-integrable Hamiltonian systems II. *Trudy Sem. Petrovsk* **5** (1979), 5-50 [Russian]. English translation: *Topics in Modern Mathematics, Petrovskii Sem. No. 5* O. A. Oleinik Ed., Consultant Bureau, New York, 1985.

[68] L. NIEDERMAN, Nonlinear stability around an equilibrium point in a Hamiltonian system. *Nonlinearity* **11** (1998), 1465-1479.

[69] L. NIEDERMAN, Hamiltonian stability and subanalytic geometry. *Ann. Inst. Fourier* **56**(3) (2006), 795-813.

[70] T. NISHIDA, A note on an existence of conditionally periodic oscillation in a one-dimensional lattice. *Mem. Fac. Engrg. Kyoto Univ.* **33** (1971), 27-34.

[71] H. POINCARÉ, *Les Méthodes Nouvelles de la Méchanique Céleste*, Vol. I-III, Gauthier-Villars, Paris, 1899.

[72] M. PETKOVŠEK, H. S. WILF, & D. ZEILBERGER, $A=B$, A. K. Peters, Wellesley, 1996.

[73] M. PETTINI, L. CASETTI, M. CERRUTI-SOLA, R. FRANZOSI, & E. G. D. COHEN, Weak and strong chaos in Fermi-Pasta-Ulam models and beyond. *Chaos* **15** (2005), 015106.

[74] J. PÖSCHEL, Integrability of Hamiltonian systems on Cantor sets. *Comm. Pure Appl. Math.* **35** (1982), 653-695.

[75] J. PÖSCHEL, Nekhoroshev estimates for quasi-convex Hamiltonian systems. *Math. Z.* **213** (1993), 187-216.

[76] J. PÖSCHEL, On Nekhoroshev's estimate at an elliptic equilibrium. *Int. Math. Res. Not.* **4** (1999), 203-215.

[77] J. PÖSCHEL, A lecture on the classical KAM theorem. *Proc. Symp. Pure Math.* **69** (2001), 707-732.

[78] P. POGGI & S. RUFFO, Exact solutions in the FPU oscillator chain. *Physica D* **103** (1997), 251-272.

[79] A. P. PRUDNIKOV, YU. A. BRYCHKOV & O. I. MARICHEV, *Integrals and Series, Vol. 1: Elementary Functions*, Gordon and Breach, New York, 1986.

[80] A. F. RANADA, A. RAMANI, B. DORIZZI, & B. GRAMMATICOS, The weak-Painlévé property as a criterion for the integrability of dynamical systems. *J. Math. Phys.* **26**(4) (1985), 708-710.

[81] B. RINK & F. VERHULST, Near-integrability of periodic FPU-chains. *Physica A* **285** (2000), 467-482.

[82] B. RINK, Symmetry and resonance in periodic FPU chains. *Comm. Math. Phys.* **218** (2001), 665-685.

[83] B. RINK, Direction reversing travelling waves in the Fermi-Pasta-Ulam chain. *J. Nonlinear Sci.* **12** (2002), 479-504.

[84] B. RINK, *Geometry and dynamics in Hamiltonian lattices*. Thesis, Universiteit Utrecht, 2003.

[85] B. RINK, Proof of Nishida's conjecture on anharmonic lattices. *Comm. Math. Phys.* **261** (2006), 613-627.

[86] H. RÜSSMANN, Invariant tori in non-degenerate nearly integrable Hamiltonian systems. *Reg. Chaot. Dyn.* **6** (2001), 119-204.

[87] V. N. SAN, On semi-global invariants for focus-focus singularities. *Topology* **42** (2003), 365-380.

[88] J. A. SANDERS, *On the theory of nonlinear resonance*, Thesis, Universiteit Utrecht, 1979.

[89] G. SÉROUL, *Programming for Mathematicians*, Springer, Berlin, 2000.

[90] M. B. SEVRYUK, The finite-dimensional reversible KAM theory. *Physica D* **112** (1998) (special issue on "Time-Reversal Symmetry in Dynamical Systems"), 132-147.

[91] S. H. STROGATZ, *Nonlinear Dynamics and Chaos: With Applications to Physics, Biology, Chemistry, and Engineering*, Addison-Wesley, Reading, 1994.

[92] G. TESCHL, Almost everything you always wanted to know about the Toda equation. *Jber. d. Dt. Math.-Verein.* **103** (2001), 149-162.

[93] M. TODA, *Theory of Nonlinear Lattices*, 2nd enl. ed., Springer Series in Solid-State Sciences **20**, Springer, Berlin, 1989.

[94] S. ULAM, *A Collection of Mathematical Problems*, Interscience Inc., New York, 1960.

[95] B. L. VAN DER WAERDEN, *Algebra I*, Heidelberger Taschenbücher, Springer, Berlin, 1966.

[96] B. C. VAN FRASSEN, *The Scientific Image*, Clarendon Press, Oxford, 1980.

[97] G. H. WALKER & J. FORD, Amplitude instability and ergodic behavior for conservative nonlinear oscillator systems. *Phys. Rev.* **188**(1) (1969), 416-432.

[98] T. P. WEISSERT, *The Genesis of Simulation in Dynamics: Pursuing the Fermi-Pasta-Ulam Problem*, Springer, New York, 1997

[99] N. J. ZABUSKY, Exact solution for the vibrations of a nonlinear continuous model string. *J. Math. Phys.* **3**(5) (1962), 1028-1039

[100] N. J. ZABUSKY, Phenomena associated with the oscillations of a nonlinear model string: The problem of Fermi, Pasta, Ulam. In: S. DROBOT & P. A. VIEBROCK (EDS.), *Mathematical Models in Physical Science*, Proceedings of a Conference at the University of Notre Dame, 1962, Prentice-Hall, Englewood Cliffs, 1962, 99-133.

[101] N. J. ZABUSKY & M. D. KRUSKAL, Interaction of "solitons" in a collisionless plasma and the recurrence of initial states. *Phys. Rev. Lett.* **15**(6) (1965), 240-243.

[102] N. J. ZABUSKY, Fermi-Pasta-Ulam, solitons and the fabric of nonlinear and computational science: History, synergetics, and visiometrics. *Chaos* **15** (2005), 015102.

[103] G. M. ZASLAVSKY, Long way from the FPU-problem to chaos. *Chaos* **15** (2005), 015103.

[104] E. ZEIDLER (HRSG.), *Teubner-Taschenbuch der Mathematik*, 2., durchgesehene Auflage. Teubner, Wiesbaden, 2003.

[105] N. T. ZUNG, Kolmogorov condition near hyperbolic singularities of integrable Hamiltonian systems. *Reg. Chaot. Dyn.* **12**(6) (2007), 680-688.

Index

action variables, 3, 48, 67, 69
action-angle coordinates, 17, 21
Arnol'd, vii

Beukers, 91
bifurcation, 19
    parameter, 9, 10, 19, 66, 71, 83
    point, 19
Birkhoff, vi, 20
    coordinates, 20
    normal form, ix, 4, 8, 20, 30, 34, 38, 49, 52, 59, 87
Boltzmann, vi

canonical
    diffeomorphism, 13
    isomorphism, 27
Cantor set, 21
celestial mechanics, 87
Chirikov, viii
combinatorics, 99
continuum limit, x, 89
convex, 4, 8, 23, 52, 59, 61, 62, 87
    m-convex, 24
Conway, 91
creative telescoping, 100
critical
    point, 9, 10, 17, 27, 71, 80
    value, 17

disjoint partition, 66

epistemological issues, xi
equipartition, vi, vii
ergodicity, vi
Euler's formula, 93

Fermi, v

fixed point, 9, 10, 17, 84
    elliptic, 9, 10, 19, 30, 68, 71, 72, 73, 78, 79, 85
        degenerate, 19, 72
    hyperbolic, 9, 10, 19, 68, 72, 76, 73, 79, 85
foliation, 9, 17, 66
Ford, vii, viii
FPU chain, vi, 1, 25, 51, 65, 87
    α-chain, 3
    β-chain, 3, 13, 52, 56, 59, 64, 65
    Dirichlet, vii, 2, 3, 7, 12, 43, 59, 88
    paradox, vii, viii, 12, 87
    periodic, vii, 1, 3, 25, 51
        even, vii, 3, 5, 9, 12, 38, 59, 88, 97
        odd, vii, 3, 12, 38, 53, 88, 97
    potential, 2
    problem, vii
frequency, 18
    map, 21

Galilei, v
Giorgilli, 24
Gustavson, viii

Hamilton, v
Hamiltonian, 15
    Dirichlet chain, 7, 8
    periodic chain, 1, 4, 5, 30, 38, 40
Hamiltonian vector field, 16, 67, 70, 83, 84
harmonic oscillator, 4, 8, 65, 70
Hartman-Grobman theorem, 20
Hénon-Heiles system, viii
Hessian, 4, 8, 51, 59

heteroclinic orbit, 9, 10, 20, 76
Hirooka, ix
homoclinic orbit, 10, 20, 78, 79
homological equation, 31, 37
Hopf
    map, 9, 67, 70
    variables, 5, 9
hyperbolic dynamics, 69

Il'yaschenko, 23
indefinite, 52, 61
index, 53, 62
integrability, 32
integrable model, vii, 12
integrable system, 7, 13, 17, 41, 59, 66, 67, 69, 83, 88
    perturbed, vi, 21
Izrailev, viii

Jackson, viii
Jacobi identity, 15
Jones, 91

KAM theorem, vii, 12, 22, 87
Kappeler, x, 3, 25, 75
KdV equation, x, 25
Kepler, v
Kolmogorov, vii
Kolmogorov's condition, 21, 62
Kronecker torus, 18
Kruskal, x
Kuhn, xii

Lagrange, v
Lax formalism, x
Leibniz rule, 15
level set, 17, 66, 68, 82
Liouville-Arnol'd-Jost theorem, 18
Lochak, 23

metastability, ix
minimal polynomial, 94
moment map, xi, 9, 10, 17, 67, 69, 88
Morbidelli, 24
Moser, vii
multiset, 35

negative definite, 63
Nekhoroshev, ix, 23
Nekhoroshev theorem, ix, 12, 23, 24, 87
Newton, v
Newton sum, 101
Newton-Girard formula, 101
Niederman, 23
Nishida, ix
nondegenerate, 4, 8, 21, 52, 53, 59, 62
    3-jet-, 24
    isoenergetically, 22, 55
nonresonant, 18, 32, 37, 91

paradigm change, xii
parametrization, 44
partition, 94
Pasta, v
Paule, 100
Poggi, 13
Poincaré, vi, 21
Poisson
    bracket, 15, 30, 41, 69, 84
    manifold, 15
    structure, 16
        nondegenerate, 16
Pöschel, x, 3, 22, 25
positive definite, 63
prime number, 93, 94

quadratic form, 63
quasiconvex, 4, 8, 23, 52, 53, 61, 62, 87
    l, m-quasiconvex, 24
    directionally, 4, 8, 24, 52, 53, 61, 62

rationally
    dependent, 18
    independent, 18
recurrence relation, 99
relative coordinates, 25
resonance, xi, 87
    lattice, 2
resonant normal form, 6, 38, 40, 49, 59, 66, 87
Riese, 100

INDEX 113

Rink, ix, 13, 43, 65, 91
Ruffo, 13
Rüssmann, 22, 55, 87

Saito, ix
San, xi, 59
Sanders, ix
sandwich theorem, 59
Schorn, 100
small divisor condition, 21
solitons, x, 13, 89
stable manifold, 20, 76, 78, 79
steepness, ix, 23, 87
strongly nonresonant frequencies, 21
symmetric polynomial, 57, 99
symmetrized coefficients, 37, 39
symplectic
    diffeomorphism, 13
    form, 16
    manifold, 16
    reduction, 18, 67, 70, 82
    structure, 16, 43, 44

thermalization, vii
Toda, ix
Toda lattice, viii, x, 3, 7, 13, 30, 54, 65, 88
torus action, 70
Tsingou, vii

Ulam, v, xi
unstable manifold, 20, 76, 78, 79

Vieta's theorem, 58

Walker, viii
Waters, viii
Weissert, vi ix, xi

Zabusky, x
Zeilberger, 100
Zung, xi, 59

Die VDM Verlagsservicegesellschaft sucht für wissenschaftliche Verlage abgeschlossene und herausragende

Dissertationen, Habilitationen, Diplomarbeiten, Master Theses, Magisterarbeiten usw.

für die kostenlose Publikation als Fachbuch.

Sie verfügen über eine Arbeit, die hohen inhaltlichen und formalen Ansprüchen genügt, und haben Interesse an einer honorarvergüteten Publikation?

Dann senden Sie bitte erste Informationen über sich und Ihre Arbeit per Email an *info@vdm-vsg.de*.

Sie erhalten kurzfristig unser Feedback!

VDM Verlagsservicegesellschaft mbH
Dudweiler Landstr. 99   Telefon +49 681 3720 174
D - 66123 Saarbrücken   Fax   +49 681 3720 1749

www.vdm-vsg.de

Die VDM Verlagsservicegesellschaft mbH vertritt

Printed by Books on Demand GmbH, Norderstedt / Germany